Sports Medicine: It's Environmental

Mark R. Hutchinson
Sérgio Rocha Piedade
Editors

Sports Medicine: It's Environmental

Springer

ISAKOS
International Society of Arthroscopy, Knee Surgery and Orthopaedic Sports Medicine

Editors
Mark R. Hutchinson
UIC Sports Medicine Center,
Department of Orthopaedics
University of Illinois at Chicago
Chicago, IL, USA

Sérgio Rocha Piedade
Exercise and Sports Medicine,
Department of Orthopedics,
Rheumatology, and Traumatology
University of Campinas – UNICAMP
Campinas, SP, Brazil

ISBN 978-3-031-84492-8 ISBN 978-3-031-84493-5 (eBook)
https://doi.org/10.1007/978-3-031-84493-5

This Springer imprint is published by the registered company Springer Nature Switzerland AG
The registered company address is: Gewerbestrasse 11, 6330 Cham, Switzerland

Preface

The primary priority for all sports medicine clinicians is the health, safety, and wellness of the athletes entrusted to their care. Most good sports clinicians are well aware of how to identify and treat the common injuries or illnesses that occur within their specific sport. The best of those clinicians are also invested in the concept of prevention that decrease the incidence of injuries and illnesses, which translate to higher continued performance of their athletes and teams.

Fundamental knowledge regarding the potential medical problems, interventions, and prevention protocols related to the specific environmental risk factors is critical to optimizing safety and care of the athletic population. *Sports Medicine: It's Environmental* targets this need and gap of knowledge by addressing the challenges, common injuries, and recommended interventions for the most common varied environments that athlete's compete in. The goal is to not only expand the knowledge and expertise of sports medicine clinicians but also optimize the health, safety, and wellness of our athletes within the environments that they compete.

Chicago, Illinois, USA

Mark R. Hutchinson

Contents

1 Introduction

Matthew A. Siegel, Sérgio Rocha Piedade,
and Mark R. Hutchinson

Victory belongs to the most tenacious is the saying registered at the Center Court of Roland Garros—**Court Philippe-Chatrier**, which means no matter how difficult the path, one should not give up on the chosen objective. It is the fuel that drives athletes; they dedicate their lives to training hard, achieving records, fighting against their body's limits and opponents, and even competing and facing different weather scenarios, such as heat, cold, altitude, wind, and so on.

Human beings have an innate desire and ambition to face challenges, whether at sea, on land, in the cold, in the heat, at altitude, or even in space. All of this has made man fight to overcome obstacles, whether losing or winning, but always go forward to build a path to fulfill a dream or a project, which is inherent to human nature. Thus, discoveries of new horizons were and are still being made, and human dreams have been materialized into achievements, most times facing the adversity of nature's forces.

In sports, environmental conditions may affect athletes' performance and health, overloading metabolic and endocrine systems and weakening the body's physiological responses to overcome adverse or extreme weather settings. It worsens when the body's thermoregulation process fails, and dehydration is associated with it.

Moreover, climate change has also created new challenges that have come up with extreme weather events: heat and cold waves, changes in rain patterns, wet and isolated (open water), UV exposure, cloudy and bad weather, mountain and altitude, and air and water pollution that are more common than before. However, defining the environment regarding sports encapsulates more than just the climate. Team physicians must also have action plans for their athletes regarding crowded mass sporting events, potential dangers, endemic and pandemic diseases, and even indoor conditions.

The team physician must face specific challenges with limited resources. This concept is akin to the practices of wilderness and military medicine. Both wilderness and military medicine were critical in developing US military front line trauma guidelines in Vietnam, depending on clinical judgment, flexibility, and creativity due to its isolation from hospital resources. In these types

M. A. Siegel (✉)
Department of Orthopaedic Surgery, University of Illinois at Chicago, Chicago, IL, USA
e-mail: msiegel2@uic.edu

S. Rocha Piedade
Exercise and Sports Medicine, Department of Orthopedics, Rheumatology, and Traumatology, University of Campinas – UNICAMP, Campinas, SP, Brazil
e-mail: piedade@unicamp.br

M. R. Hutchinson
UIC Sports Medicine Center, Department of Orthopaedics, University of Illinois at Chicago, Chicago, IL, USA

M. R. Hutchinson, S. Rocha Piedade (eds.), *Sports Medicine: It's Environmental*,
https://doi.org/10.1007/978-3-031-84493-5_1

of medicine, improvisation is tapped when using duct tape, knives, and netting to stabilize patients, even though these items may be thought of as "nonmedical" [1, 2]. Subsequently, protocols were developed for the common citizen [1]. Indeed, even in recent times at the highest level of sport, medical leadership should take advantage of the expertise of military and wilderness medicine experts. Adapted from the Marine Corps Marathon: Medical Triage and Algorithms [3], the US Olympic and Paralympic Committee (USOPC) created their own protocols that were utilized to save a life during the 2020 Tokyo Olympic Games [4] (Fig. 1.1).

The US high-performance center head team physician, Dr. Mark Hutchinson, and the high-performance center emergency response team were alerted that there was a person collapsed. The emergency response team sprang into action driving their fully supplied golf cart to the scene. Found down was not an athlete, but an elderly male gardener who was working inside of the fenced off Olympic Grounds area, which was restricted due to COVID-19 precautions. The patient with unknown medical history was unresponsive for an unknown amount of time. An emergency medicine trained and a Team USA Judo physician, Dr. Christina Yannetsos, immediately initiated the cardiac care algorithm. It was a sweltering hot summer Olympic day, and the patient was swiftly diagnosed with heat-related illness (It's Hot). Dr. Hutchinson and the ancillary medical staff simultaneously commenced the hyperthermia protocol alongside the cardiac code (Fig. 1.1). The patient's clothing was removed, and prechilled water-soaked towels were taken from the golf cart and applied for rapid external cooling. In a true team effort, the patient became responsive, his vital signs stabilized, he was transferred to pre-identified indoor cooling area, and he recovered well with further monitoring. Optimizing your medical team with the appropriate personnel will allow for the best chance at success [5]. This case did not include a collapsed athlete but rather a civilian who was likely not heat acclimated [6]. This anecdote helps to emphasize the importance of being prepared for a broad range of challenges when facing environmental challenges and underscores the need for the emergency response teams and team physicians to have practiced well-established protocols. Having standardized protocols in place

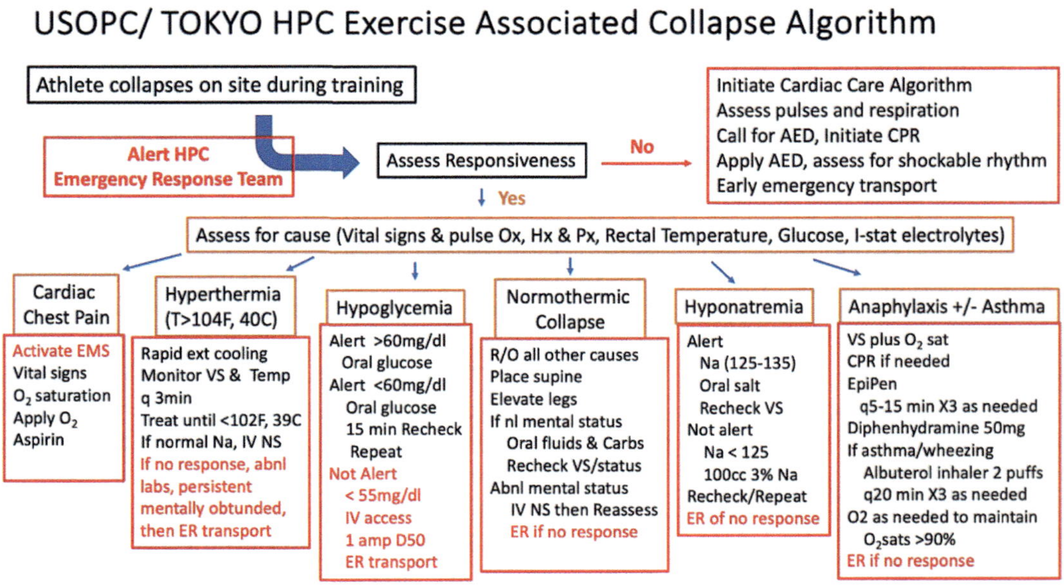

Fig. 1.1 US Olympic and Paralympic Committee (USOPC) high-performance center (HPC) exercise-associated collapse algorithm used in the 2020 Tokyo Olympics. (Adapted from the Marine Corps Marathon: Medical Triage and Algorithms)

allowed the best patient care to be administered without wasting precious time, even in the face of the COVID-19 pandemic.

1.1 Preparedness

Coaches will harp on their players about how important it is to prepare for their competition. This may be mentally, physically, and/or emotionally. To effectively execute a game plan, the players and/or team must be prepared. Preparedness is also the crux of being a successful team physician. And just as the athletes will look to their coaches in the game's dire, strategic moments, the athletes will trust their team physician to know the environment to keep them safe and healthy. Most of the potential environmental roadblocks can be avoided prior to the game even starting. Does the sporting venue have the correct temperature, humidity levels, lighting, and surface conditions? (It's Indoor). Do you have a plan in place when the open water competition's environment changes and suddenly has strong currents, high waves, and hazardous marine life? (It's Wet and Isolated). What if the UV index is 9 and your athlete develops a severe sunburn? (It's Sunny and Bright). Military medicine has been analyzing and adapting to preventable roadblocks for decades and has since learned how to minimize preventable deaths that includes hemorrhage, tension pneumothorax, airway obstructions, and more [2, 7]. Prevention is a common theme in wilderness medicine as well [6, 8, 9]. Moreover, wilderness medicine has many protocols that can also be used by the team physician, such as those to treat frostbite of the fingers (It's Cold). This includes supportive measures, such as encouraging exercise, changing gloves to mitts for increased insulation, and drying the hand, if wet, or rapid rewarming [8, 10]. The prepared team physician will intricately plan to overcome the unique challenges in certain athletes, communication, as well as extrication and access to athletes, access to local emergency health systems, access to emergency transport, and access to necessary equipment.

1.2 The Athlete

Who is the athlete? Do they need a specific type of care or attention? Now, imagine you are covering the Paralympics, and there is an excessive heat warning during the gold medal event. Some of the athletes have traumatic bilateral lower extremity amputations, but others have a similar condition, yet it is congenital. The patient with the congenital deformity may have other medically complex issues you must take into account. Still, both athletes are at higher risk for heat exhaustion and heat stroke due to having less surface area that can dissipate the heat. Symptoms of dehydration may also be more pronounced due to less circulating blood volume. Do you have designated protocols in place to account for these issues? If a Paralympic wheelchair basketball athlete presented to you with shoulder pain, would you intervene quicker? A wheelchair-dependent Paralympic athlete is dependent on their upper extremity for performance, and this pain can quickly affect the athlete's mobility. Treatment may require more deliberate and timely intervention with NSAIDs and other modalities that you may not normally initiate with an able-bodied athlete (Fig. 1.2). If you are treating athletes with spinal cord trauma, and they have an ulcer/lesion, or are experiencing

Fig. 1.2 A free-throw attempt by Team Australia at the 2000 Sydney Paralympics. (Reproduced from Wikimedia Commons at https://commons.wikimedia.org/wiki/File:Wheelchair_basketball_Nick_Morris_shoots_-_3b_-_Sydney_match_photo_2000-10-25.jpg by Sport the library/Bill Bachman; Australian Paralympic Committee Licensed under Creative Commons Attribution-ShareAlike 3.0 Unported (CC BY-SA 3.0))

bladder retention leading to severe hypertension, what measures do you have to prevent or treat stroke-like symptoms? Do you have access to any unique assessment tools or equipment necessary to optimize your care of a Paralympic athlete? For example, if you have a sit skier who was injured and needed help down the slope, you will more likely need a sled to best assist your athlete.

1.3 Communication

Is your cell phone charged? What if you are covering a mountain biking event and do not have cell service? What if this is an international event, and you do not speak the native language? Do you have access to an interpreter? Is the patient not communicating because of the language barrier? Or did they sustain head trauma? Or are they having an asthma exacerbation due to the poor air quality? (It's Polluted). Do you and your colleagues have a team lead and chain of command for emergency situations? Have you practiced the protocols and checked communication pathways before the start of the event? The US military medicine operates with five levels of escalating care with clear roles, which mirrors its combat military's organizational structure [2, 11]. In the army, combat medics initiate first responder care at Level 1, typically in infantry companies. Companies are composed of multiple platoons and squads, and the medic may join any of these teams as needed to help complete missions. The entire military medicine organization depends on the ability of these medics to initiate treatment, triage appropriately, and communicate with Level II care [11]. The battalion aid station (composed of multiple companies) is the facility where primary care and emergency medicine physicians first treat injured soldiers from multiple companies. Brigades (composed of multiple battalions) have Charlie Meds that provide Level 2 care. These are similar to the battalion aid stations but have more sophisticated resources, such as imaging capabilities. A Level 2 Charlie Med may also have a forward surgical team (considered Level 2+), which has some ability for stabilizing surgery within the combat area; this would include such things as damage control orthopedics [2, 11]. This Level 2+ team may include "one orthopedic surgeon, three general surgeons, two nurse anesthetists, one critical care nurse, and additional nursing staff" [11]. Depending on the acuity of care required, military medicine requires its providers to have the skillset to triage patients and bypass Level 2 entirely. If a patient is stable and needs surgery, it would be favored to transfer the patient directly to a Level 3 Combat Support Hospital within a division (composed of multiple brigades). Level 4 care commences away from direct combat and consists of Enhanced Combat Support Hospitals. This is initiated at the corps level (composed of multiple divisions) and higher [2]. Finally, military medical treatment inside of the continental USA is considered Level 5 care [2, 11]. This is a vastly complex system with multiple teams dedicated to transportation, supply management, and other logistics that must seamlessly communicate together for life and limb saving treatments [2, 11]. Care of athletes in various environments should have a similar, if not as intense, concept of who is the first responder, what are their limits and when should they transfer to a higher level of care, and how via communication and type of transportation will the athlete receive the best and highest quality care in a timely fashion. As the team physician, do you have the infrastructure necessary to best treat your athletes? Can your team communicate through verbal and nonverbal communication? You must also be mindful with how you are communicating with your team and your athlete. A secondary role of the team physician provides a unique communication avenue to act as a mentor to aspiring young athletes.

1.4 Equipment

Environmental issues can introduce unique challenges regarding access to necessary medical equipment. Who is bringing what? Do you have appropriate extrication device(s)? Where is it placed in relationship to the playing surface? Is it on an adjacent boat in open water events?

(It's Wet). Are multiple packages of emergency response equipment stationed at various stages along an extended outdoor course? (It's Rugged and Isolated). Is the cervical collar in your personal kit or the team bag on site? Where is the nearest automated early defibrillation [12]? Who is carrying and where are the splints? Are they small and disposable, or are they prefabricated splints? Where is the stretcher/backboard? Some of this equipment, and certainly the collection of all of it, can be extremely bulky and may be difficult to carry in your own backpack kit. Having it stationed at known access points or assigning someone to retrieve and deliver to the site of injury is usually the most efficient plan. Is there a dedicated staff member responsible for supply logistics? Assuring that access is straightforward and without barriers is critical to success. Is all your personal equipment prepared and ready? Is your cell phone charged? Do you have extra batteries and chargers for flashlights, AED, or powered equipment? In military medicine, the logistics of the immediate restocking of medical equipment has its own dedicated team that is separate from other military needs, such as ammunition and food [2]. Similarly, it is important to have a mechanism of communication to document what was used so that it can be replaced and that the equipment can be ready for the next event.

1.5 Extrication and Access to Athletes

Access to a downed athlete and the specific plan and means of extrication is a critical facet when addressing any environmental challenge. Prior to rushing to your injured athlete, you should ask yourself, "Is the scene safe for you to proceed" [1]? (It's Dangerous). If your mountain cycling athlete is down, can you get to the athlete safely due to the terrain and altitude? (It's Rugged and Isolated). Prior to beginning the extrication, do you know if you even have access to the athletes? Who has first responder access? At every international and multisport event, a local provider should have a credential that allows immediate

Fig. 1.3 Cycling crash at the 2008 Tour de Taiwan. (Reproduced from Wikimedia Commons at https://commons.wikimedia.org/wiki/File:2008TourDeTaiwan_Stage7_2nd_crash.jpg by Astrokey44 under the GNU Free Documentation License and the Creative Commons Attribution-Share Alike Attribution-Share Alike 4.0 International, 3.0 Unported, 2.5 Generic, 2.0 Generic and 1.0 Generic License with attribution to the Rico Shen)

on-field access. However, you as your team's physician may not. Do you have to wait in an off-site area for them to provide initial on-field management? Can you get onto the barrier-protected course if there is a bicycle collision (Fig. 1.3)?

In the 2020 Tokyo Olympic Games, a Team USA athlete was involved in a cycling crash. The host country provided the emergency response team, and their doctors were the only physicians with immediate access to on-field care, and Team USA was forced to meet them at the ambulance. The host emergency responders initiated their emergency response plan which included boarding and immobilization but not the application of a cervical collar. The lack of a cervical collar was not a failure of their planned protocol as it was not standard practice in the host country. However, this would have been inconsistent if the event had occurred in the USA. In the case anecdote, a cervical collar was ultimately applied by the athlete's team physician when they were able to reconnect in the ambulance. The message here is NOT to be critical of the host country but rather to emphasize the importance of communication between a host country or organizer and the athlete's medical team. Communication should occur before the event to understand the protocols, at the event to understand who has access and is responsible, and after the event to understand who assumes responsibility after extrication. Different places, countries, and organizations may have different rules for access and protocols for on-field injuries, which you may or may not agree with. Especially at international events, protocols need to be discussed preceding any injury. The same principles ring true when you are on the sideline of your local high school football team.

Another example highlighting the importance of ironing out protocols occurred at the Lake Placid 2023 Winter World University Games, when a hockey player suffered a shattering check into the boards. The athlete sustained head and neck trauma which alerted the medical team for the potential of paralysis. The athlete landed face down on the ice wedged between the ice and the barrier wall. The initial rescue response was flawless—the team physician and medical team hurried onto the ice with special attachments to their shoes to provide traction, the athlete's cervical spine was stabilized as the player was flipped supine, a cervical collar was applied, and the athlete was taped onto a hard backboard which was transferred to the wheel stretcher. The athlete was safely taken off the ice; however, based on regional and state guidelines, the local emergency medical technicians wanted to again transfer the athlete off the hard spine board onto the cushioned stretcher prior to transport to the nearest hospital. The basis of the regional guidelines was that the hard spine board was uncomfortable and increased the risk of skin problems if extended transport was required. Due to the elevated risk of cervical spine injury, the team physician took a hard opposing stance against this, especially because the receiving hospital was only seventeen minutes away. In the scenario provided, the local emergency medical technicians deferred to the physician; however, the lesson to be learned was that the protocol and plan for each step of the extrication would have best been planned and understood well before the event in question.

Understanding and preplanning of both the extrication plan and access of the medical team is critical to optimizing outcomes in any environment. You are now covering Women's Gymnastics in the 2016 Rio Olympic Games. You were given an infinity access pass that allows you to have access to all facilities. With this pass, you had universal access, right? Wrong. During various competitions including gymnastics and equestrian, certain venues may require additional security credentials. As this was a highly televised event, the governing organization did not want extra people, including each country's team physician, on the floor and on camera. In equestrian events and in an effort to restrict disease transmission, international horses may have a restricted area that requires additional clearance for access. (It's Contagious). What if you need to evaluate your athlete in this area? Conversations and negotiations with the regulating powers and governing bodies need to take place long before the competition commences in order to balance their needs but also provide efficient access to optimize care of the athlete. The solution is usually achieved with communication and a sincere respect for each party's needs and wishes. Local providers do not want the ultimate responsibility of your star athlete, yet you need to respect their need for both appearance, safety, and security.

1.6 Evacuation

Now that you have treated and stabilized your patient, you may need to proceed with escalation of care and evacuation. Where is the closest hospital? Do you have emergency medical services and transport in place? Are you taking an ambulance or a helicopter? How does the environment or weather impact these choices? (It's Stormy). Are these services on call or immediately available? To get to these services, do you need to move through a large amount of people? (It's Crowded). Who is your liaison? Is there a language barrier? Thankfully, sports-related injuries are rarely a life or death situation; nonetheless, being prepared and having established protocols similar to transporting from the battlefield or from the wilderness is important [1, 2].

Most commonly after providing Level 1 and sometimes Level 2 care, an athletic event will take advantage of the local emergency health systems for the next level of care. Pre-event communication should include notification to the hospital and health systems of the event to assure their awareness and so that they can staff appropriately. Ideally, a specific liaison or contact person should be identified within the emergency room, hospital, and/or healthcare system to optimize access and efficiency once an event occurs. The contact person can be the triage nurse, on-call physician, or hospital administrator who is empowered to expedite care and activate the system. The team physician must have this all figured out prior to the event to have a quick and smooth transition when transferring care.

The above themes are intimately intertwined but must be individually considered. One misstep in the preparation can lead to significant delays of the medical response which in turn risks potential harm to the athlete. Even if your local team has its hospital across the street, the necessary equipment must be present at the competition. Your protocols must be well established to safely stabilize and transport your injured athletes. Now, imagine you are covering an international skiing event in the 2018 Winter Olympics in Pyeongchang, South Korea. One of the athletes is down in the middle of the slope and clearly injured. The only way to get access to the athlete is by skiing down the Olympic-grade mountain to the access point, which is halfway down the slope. Fortunately, you have been cleared to have direct on-field access. However, are you able to ski? Do you have appropriately fitting skis available? This situation may loosely parallel wilderness medicine evacuation plans, such as an avalanche rescue or any type of high- to low-angle rope rescue [1]. The athlete does not speak your language. Is your cell phone working? Do you have an interpreter with you? Can the interpreter ski? The athlete's condition is severe. The skier has sustained an open tibia and fibula fracture and has a limb threatening injury. Who has the splint supplies to stabilize the leg? Transport to the nearest major hospital is required, but this is one hour away. Can you get IV access and administer antibiotics or pain medications before arriving to the hospital? The initial plan was to have helicopters airlift the athlete quickly and safely. But there is a problem—the weather conditions are not safe for helicopter flight. The backup extrication plan involved ambulances driving down the mountain, which was the only option at this juncture. While on the way to the hospital in the ambulance, you notify the emergency department and the on-call physician whom you have already created a relationship with. Ultimately, the athlete was treated with temporary fixation and definitively fixed after transport home. This anecdote, based on a real-life scenario, serves to emphasize the point that the environment can pose unique challenges when caring for athletes which can be minimized by thorough pre-competition planning.

Lastly, your environmental challenges may present you with ethical dilemmas. It may be inconvenient to get the appropriate access or to the intended place of care. You may be faced with a time delay and pressures from the athletes, coaches, and fans to have the athlete return to play. If medical care is indicated, the team physician has the responsibility to do what is right. In the above example, if the skier with the fracture has questionable neurovascular symptoms and the hospital is one hour away, you need to decide if you should perform a reduction or splint in situ prior to imaging and workup. In all types of sporting events, the team physician must be mindful to protect the athlete from fans and other interested third parties. The team physician-patient care relationship must be kept confidential.

All these and other challenging scenarios in sports practice have triggered Dr. Mark Hutchinson and Dr. Sérgio Rocha Piedade to organize this book, presenting a comprehensive understanding of the physiological and metabolic responses of the athlete's body all while facing adverse environmental conditions. Further, this textbook fulfills the need to have a central go-to reference with clear guidelines for how to respond to environmental conditions with variable responses, such as lightening (It's Stormy). Being a team physician is an honor. All team physicians should aspire to organize and implement a system that can rapidly respond to any environmental challenge, parallel to the impressive ability the military has to mobilize its medical care [2]. Just as wilderness medicine was developed by three doctors in 1983 and has developed through several editions in a primary textbook [13, 14], this textbook too can serve as the first foundation of accepted environmental practice for the team physician.

1.7 Key Takeaways

- The successful team physician is the one that is prepared as well as able to adapt to spontaneous environmental challenges. Creativity and resourcefulness must be utilized when limited resources and environmental obstacles are encountered.
- Understand your athletes' unique needs. Have all protocols and required supplies available in advance.
- Make sure you have the appropriate access to best care for your athlete on the field and on the sideline and, if necessary, have the right access/contacts to escalate care to a higher level (e.g., hospital). Negotiations and clearances for obtaining the correct access need to happen before the event begins.
- It is vital to have emergency and return to play protocols approved prior to a real-life scenario.

References

1. Sholl JM, Curcio EP. An introduction to wilderness medicine. Emerg Med Clin North Am. 2004;22(2):265–79.
2. Hetz SP. Introduction to military medicine: a brief overview. Surg Clin N Am. 2006;86(3):675–88.
3. O'Connor F, Adams WB, Madsen CM, Beutler AI, Brennan Jr. FH, DeLuca J, et al. Managing the collapsed runner: marine corps Marathon medical triage and algorithms [internet]. CHAMP, THE MARINE CORPS MARATHON AND THE INTERNATIONAL INSTITUTE FOR RACE MEDICINE; 2020. Available from: https://champ.usuhs.edu/sites/default/files/2021-06/MCM%20 2020%20ALGORITHM%20HANDBOOK%20 Final.pdf
4. Hutchinson M, Finnoff J. USOPC/TOKYO HPC exercise associated collapse algorithm. United States Olympic and Paralympic Committee; 2020.
5. Chung KK. Optimizing military medicine: a team approach. Mil Med. 2021;186(Supplement_3):51–2.
6. Lipman GS, Gaudio FG, Eifling KP, Ellis MA, Otten EM, Grissom CK. Wilderness medical society clinical practice guidelines for the prevention and treatment of heat illness: 2019 update. Wilderness Environ Med. 2019;30(4_suppl):S33–46.
7. Bellamy RF. The causes of death in conventional land warfare: implications for combat casualty care research. Mil Med. 1984;149(2):55–62.
8. McIntosh SE, Freer L, Grissom CK, Auerbach PS, Rodway GW, Cochran A, et al. Wilderness medical society clinical practice guidelines for the prevention and treatment of frostbite: 2019 update. Wilderness Environ Med. 2019;30(4_suppl):S19–32.
9. Goodman T, Iserson KV, Strich H. Wilderness mortalities: a 13-year experience. Ann Emerg Med. 2001;37(3):279–83.
10. Hall A, Sexton J, Lynch B, Boecker F, Davis EP, Sturgill E, et al. Frostbite and immersion foot care. Mil Med. 2018;183(suppl_2):168–71.
11. Bagg MR, Covey DC, Powell ETI. Levels of medical Care in the Global war on terrorism. JAAOS J Am Acad Orthop Surg. 2006;14(10):S7.
12. Cusano A, Inclan PM, Jackson T, Weiss LJ, Barnes RP, Kinderknecht JJ, et al. Team approach: diagnosis, management, and prevention of sudden cardiac arrest in the athlete. JBJS Rev. 2024;12(3):e23.00225.
13. Sward DG, Bennett BL. Wilderness medicine. World J Emerg Med. 2014;5(1):5–15.
14. Auerbach PS. Wilderness medicine. 6th ed. Philadelphia: Elsevier/Mosby; 2012. 2277 p

It's Hot

2

Sérgio Rocha Piedade,
Leonardo Augusto de Souza Beck,
and Daniel Miranda Ferreira

2.1 Introduction

Athletes and regular sports practitioners are constantly attracted and motivated to seek new goals and personal achievements and push their body limits, even when it is hot! Although warmer and sunny days may, in principle, be an invitation to practice sports, we must always be aware of the adverse effects of heat on our body, especially when temperatures are higher. Warmer and sunny days may, in principle, be an invitation to practice sports, but we must always be aware of the adverse effects of heat on our body, especially when temperatures are higher (Fig. 2.1). Hot ambiance may have an inherent risk of heat illness secondary to body hyperthermia, even in highly active individuals or athletes [1, 2].

Heat stress results from the body's metabolism, muscle exercise, and environmental sources, and the body's thermoregulation control depends on the efficiency of balancing heat loss and gain to maintain the core temperature at baseline values. More than that, heat stress is critical in athletes' performance, negatively affecting temperature regulation; body fluid balance; cardiovascular, metabolic, and neuromuscular physiological responses; and thermoregulatory control, which can cause inefficient heat illness [3]. Generally, for normally healthy and active individuals, exertional heat illness will be likely to take place when high-intense exercises are performed over long periods. Endurance sports, like marathons performed at hot temperatures, may cause hyperthermia, neurological impairment, and muscle fatigue due to the substantial physical demands on the athletes [4]. Although hot weather strain impairs athletes' capacity to enhance endurance exercise and sports performance, acclimatization training could improve athletes' ability to face these adverse conditions [5, 6]. This chapter, It's Hot!, discusses physiological aspects of human temperature, the body's thermoregulation, the harmful effect of heat combined with dehydration in sports performance, the importance of euhydration, and the diagnosis, predisposing factors, treatment, and prevention measures of clinical presentations of exertional heat-related illness.

S. Rocha Piedade (✉)
Exercise and Sports Medicine, Department of Orthopedics, Rheumatology, and Traumatology, University of Campinas – UNICAMP, Campinas, SP, Brazil
e-mail: piedade@unicamp.br

L. Augusto de Souza Beck
Exercise and Sports Medicine, Department of Radiology, University of Campinas – UNICAMP, Campinas, SP, Brazil

D. Miranda Ferreira
Exercise and Sports Medicine, Department of Radiology, University of Campinas – UNICAMP, Campinas, SP, Brazil

Faculty of Medicine, São Leopoldo Mandic, Campinas, SP, Brazil

M. R. Hutchinson, S. Rocha Piedade (eds.), *Sports Medicine: It's Environmental*,
https://doi.org/10.1007/978-3-031-84493-5_2

Fig. 2.1 Sports practitioner running on the beach on a hot sunny day at Jatiuca Beach in Maceio, Brazil. (Courtesy Ana Karina Piedade)

2.2 Case Anecdote/Challenge

Running is, by far, an exciting physical activity that triggers people's passion for sports and challenges their body limits. This anecdote is about an amateur runner who had trained for six months before participating in his first marathon. During his training, he had minor muscle injuries that healed but somehow did not allow him to condition himself better to face the environmental conditions commonly related to this race—the heat and humidity. On the day of the race, the event atmosphere was terrific, and he was highly motivated (with a "warrior mentality") to complete the 42.195 km course.

His performance was OK in the first half of the run, but after 27 km, the environment temperature rose to 28 °C, the humidity was around 60, and he started becoming exhausted. He felt that his body was excessively hot and realized that his body's thermoregulation was not working efficiently. The clinical symptoms worsened progressively: he felt dizzy, fatigued, and confused. At 31 km, he had seizures, lost consciousness, and collapsed. Fortunately, the medical staff promptly diagnosed the heat stroke and started his emergency care. Considering this scenario, some questions come up:

- Which preventive measures should an athlete take into account for heat-related illnesses in sports?
- What are the mechanisms involved in the human body's thermoregulation?
- What are the body's physiological responses following heat environment acclimatization?
- How should the athlete suffering a heat stroke be rescued?
- When should athlete's body cooling be started? How should cooling be done?

2.3 Epidemiology of Heat-Related Illness

Exertional heat-related illness is a clinical condition resulting from hyperthermia that impairs the body's thermoregulation; heat stroke is the most extreme clinical condition [7, 8]. *Due to inconsistency in registering the cases of exertional heat stroke, terminology, and diagnosis, the exact prevalence level is uncertain* [9]. However, it is prevalent in competitive sports and the military, and sports such as long-distance and endurance running, hockey field, American football, lacrosse, soccer, and so on have been implicated. On the other hand, elite athletics has shown a low incidence of heat stroke [9–11].

However, extrinsic and intrinsic factors play a crucial role in triggering this dangerous condition, such as environmental conditions (heat and elevated humidity), athlete's health condition (age, individual heat resilience, level of hydration, medication, and drug use), and training conditions (lack of acclimatization) [9–11]. In addition, in some sports and the military, the equipment (uniforms, helmets, protective clothes, and so on) may also create difficulties in the body's heat dissipation, and therefore, the physician must be aware of that.

In the literature, the incidence of heat-related illness in sports is challenging because most data recording is not standardized, approaching only a specific sports modality. Moreover, physical demands vary according to each sport practiced worldwide, making it difficult to define the actual incidence of heat-related illness and related sports [11, 12]. In this context, the Summer Olympic Games could work as a "global lab" where athletes from different sports compete in the same environmental scenario [12]. Therefore, it is a fantastic opportunity to register data and assess the incidence of heat-related illness more carefully, comparing it with different sports modalities. Moreover, it is an authentic global melting pot, an event bringing together athletes from various countries, cultures, phenotypes, etc., which could contribute to a comprehensive analysis of heat-related illness and the role of physical demands of each sport on it. Tanaka et al. [13] reported an incidence rate of heat stroke of 3.40% in the marathon and walk race, 0.29% in tennis, and 0.13% in track and field events during the Tokyo 2020 Olympic and Paralympic Games.

2.4 Pathophysiology of Heat-Related Illness: Human Body Temperature

Humans are homeotherms, so the body's temperature is relatively stable. The normal temperature of the human body is 37 °C, but during the day, it may vary by 0.5 °C due to metabolic and hormone processes, level of physical activity, etc. [14, 15]. This temperature range is adequate for the body's metabolism, enzyme activity, and functioning of standard body systems and organs [16].

It is essential to state that the temperature is not uniform for all body regions; it is higher and more constant in the core body, while in the skin, it is lower and variable due to external influences, such as environmental conditions [17]. Moreover, body temperature may suffer endogenous and exogenous actions, jeopardizing the function of human biological systems [5, 18].

The human body can generally be divided into two compartments concerning temperature: the core and the shell (i.e., skin and peripheral tissues). Although the muscle's temperature can also be considered a third compartment, contributing to a more accurate mean body temperature analysis, it is an invasive procedure, and measuring it is not used routinely [19].

As pointed out, the skin temperature is not regularly stable, as it is influenced by thermal environmental conditions, acting as an interface to them. In general, the skin temperature can be classified as cool at <30 °C, warm from 30 to 34.9 °C, and as hot when >34 °C [20, 21]. Therefore, body temperature is one of the critical parameters for assessing vital signs of health status [22].

2.4.1 Mechanisms of Thermoregulation

Thermogenesis is the process of heat production in the organism, resulting from the energy produced by basal metabolism, muscle exercise, and associated metabolisms. It strengthens the vital role that the body's thermoregulation plays, as it influences the physiological homeostasis, structure, and function of biological tissues, such as plasma membrane, protein configuration, enzymatic catalysis, etc. [14, 16]. Thus, thermoregulation is the physiologic capability of the body to maintain its temperature within certain boundaries when facing adverse environmental condi-

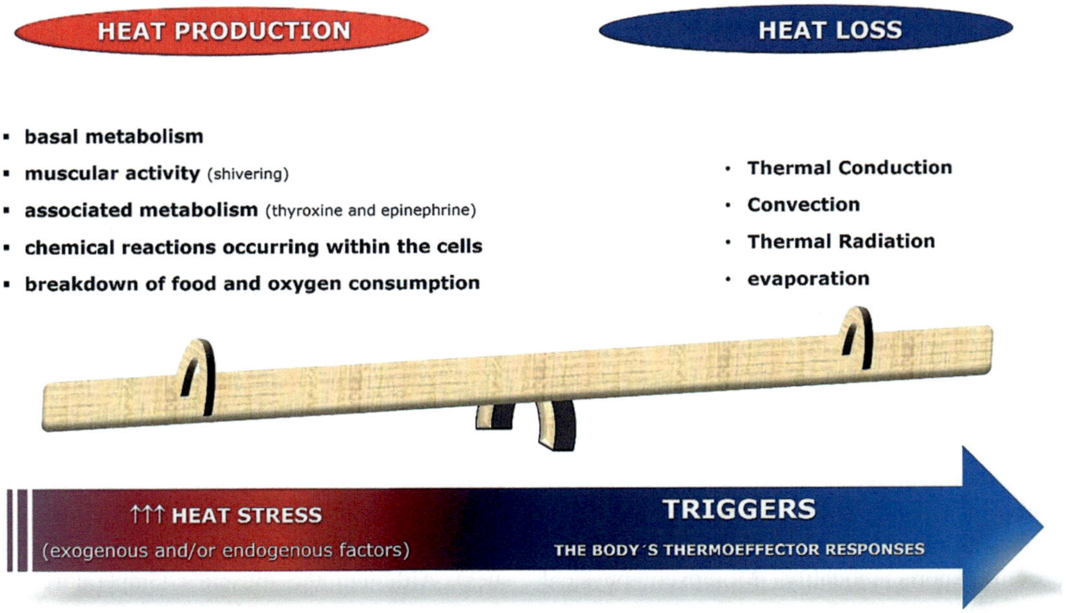

Fig. 2.2 Conditions related to body heat production and loss. (Courtesy Ana Karina Piedade)

tions [10, 18]. Figure 2.2 summarizes the conditions related to the body's heat production and loss, when the increase of core temperature due to endogenous and exogenous factors will stimulate the thermoeffector responses to maintain the core temperature within safe limits.

Humans' tolerance to cold and hot environments is not similar: the body may tolerate a 10° temperature decline but only an increase of 5°in the normal body temperature. An example is running at 5°–10 °C external temperature, commonly related to better athlete performance than at hotter weather temperatures [23]. Although hot weather strain may impair athletes' capacity to enhance endurance exercise and sports performance, acclimatization training may improve athletes' ability to face these adverse conditions [5, 6], and individual variability should also be taken into account [5, 18]. In both conditions, the body displays specific thermoregulation mechanisms, where the hypothalamus acts as a thermostat that triggers mechanisms to achieve and maintain the body's thermal regulation, guided by the skin and peripheral tissues (shell) and central (hypothalamus) thermoreceptors [20, 21].

The body's physiologic responses can reverse conditions involving slight variations in the body temperature [5]. On the other hand, extreme variations in body temperature can have more harmful consequences and may cause organ failure, coma, or even death [24–26]. Thus, thermoregulation could be achieved within physiological limits when environmental conditions change. It involves metabolic, endocrine, and physiological mechanisms guaranteeing that organs and systems work regularly. The homeostatic process of thermoregulation is critical to balance heat production and loss, keeping the body's temperature at baseline values and ensuring proper conditions for the enzymes and immunological responses to work suitably [5, 16, 22].

The brain, skin, skeletal muscles, sweat glands, and vascular, endocrine, and nervous systems are prioritized in the thermoregulation process. The skin receptors and the blood flow to the sensors of the hypothalamus "thermostat" (central control area) act as the primary trigger mechanisms for controlling body heat. The hypothalamus has two thermoregulatory regions, where the anterior is implicated with heat loss, while the posterior one is involved with the mechanism of increasing heat production and gain [20, 27].

2.4.2 The Thermal Gradient Between the Skin and the Environment

The body's metabolism generates energy, fueling our muscle body to work, but it also releases heat. Moreover, environmental conditions (humidity, temperature, air velocity, solar radiation), athlete's intrinsic features (age, sex, body morphology, and mass), and worn outfit could also cause an increase in the body temperature. All these conditions can display different mechanisms of heat dissipation to maintain the body temperature balanced [14, 28].

In thermodynamics, the gradient of heat transferring flows from a hotter place to a less hot one. Consequently, when the skin temperature is higher than the environmental temperature, the *gradient of thermal transfer is positive*, and the excess heat is transferred to the environment, preventing the body from overheating. Inversely, environmental temperatures exceeding the body's temperature result in a *negative gradient of thermal transfer*, and the heat from the environment is transferred to the body, increasing body temperature [10, 27].

The human body thermoregulation has four thermal mechanisms for heat loss: dry heat flow (conduction, radiation, and convection) and evaporative heat flow (Fig. 2.3). *Thermal Conduction* involves heat transfer through direct contact, occurring when the body's surface is in contact with a nearby solid surface. *In thermal convection*, a heat exchange takes place when a fluid medium surrounds the body's surface, moving away from the skin, resulting in heat transfer from the skin to the air, potentialized in case of high airspeed. *In thermal radiation,* heat exchange is done by electromagnetic waves, transferring energy from the body's surface to the external environment and making the skin lose heat to the ambient temperature. When exposed to the sun, the skin absorbs the heat [10].

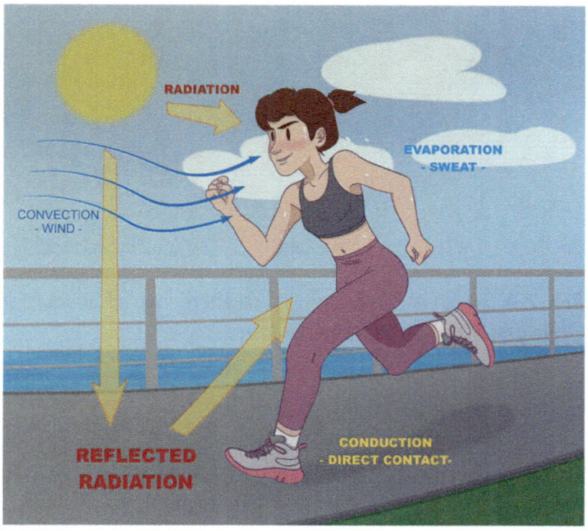

Fig. 2.3 Mechanisms of the human body's thermoregulation heat exchange from the body to the surrounding environment: (**convection**, movement of fluids; **conduction**, direct contact; **radiation**, electromagnetic waves; and **evaporation**, heat loss by sweat). (Design—Courtesy Marina Percario Piedade)

Thermal evaporation is the most efficient mechanism for our body to lose heat to the environment. It may occur in different manners: water vapor lost through the lungs, diffusion of water through the superficial layers of the skin (perspiration), and mainly *sweat evaporation* (sweating), potentialized by the wind action. However, when the relative humidity is elevated, i.e., above 60, sweating is jeopardized, and the heat exchange is less efficient, creating *a more uncomfortable environment for sports practice and exercising* [28, 29]. Moreover, the larger body mass and body surface may interfere with heat transferring from the body to the environment and how our temperature sensors understand the air temperature (thermal sensation) [30].

2.4.3 Exposed Body Surface Area and Body's Heat Transfer

The body surface area is a critical point in the body's heat loss, which occurs proportionally to the gradient between the skin and the external environment and the available body surface to heat exchange [10, 27].

Fig. 2.4 (**a–c**) A driver being refreshed with a bucket of ice water thrown by their team during a car race. (Storyboard—Courtesy Marina Percario Piedade)

Moreover, the type of sports clothing could also play a vital role in the body's heat transfer gradient. An anecdotal example of this issue is reported in car racing back in the 1960s. Due to high temperatures inside the car, the driver used to adopt the strategy of stopping their car during a moment in the race to be refreshed by their team, who threw a bucket of ice water on the driver sitting in the racing car (Fig. 2.4a–c).

2.4.4 Heat and Dehydration

2.4.4.1 The Impact of Climate Change

Moreover, climate change has also resulted in more hot days and heat waves presenting an additional factor to be taken into account in determining the best hydration strategy to preserve athlete's health [31, 32]. In addition, several intrinsic factors are also involved in the body's thermoregulation process, such as age, gender, athlete's previous injuries, and comorbidities and, therefore, should be assessed [33].

2.4.4.2 Thermoregulation and Hydration

The thermoregulation process increases the body's physiological strain at extreme heat or even heat wave conditions—the maintenance of this condition possibly leads to dehydration, mainly by prolonged endurance exercises [20, 24]. Heat stress harms aerobic performance, and when associated with dehydration, it also increases the body's thermal and cardiovascular strain. The maintenance of this clinical status could result in severe disorders due to the dehydration process and cell damage, such as seizures, swelling of the brain, kidney failure, shock, coma, and even death. It occurs when the body responds inefficiently to cool down and maintain a healthy temperature.

It reinforces that maintaining adequate and balanced fluid and sodium levels is crucial to optimizing sports performance, preventing metabolic strain and consequently acting to avoid the above clinical conditions. However, the sports physician should be aware that thirst perception is only triggered when the plasmatic osmolarity rises from 1% to 2%. Therefore, thirst is not an efficient guide to assess the dehydration status or the need for fluid repositioning during exertion [34, 35].

2.4.4.3 The Role of Euhydration

During sports and physical exercises, the body loses fluid, a physiologic thermoregulation response which occurs mainly through sweating, metabolic water, and respiration and also helps promote heat dissipation from the body to the surrounding environment. However, vigorous endurance exercises and sports at hot temperatures may cause a considerable depletion of fluid balance, resulting in dehydration. Progressively, this adverse condition triggers a cascade of events comprising increased plasma osmolarity, reduced heat loss and sweating rate (hypohidrosis), and impaired cardiac filling, reducing the athlete's physical capacity and predisposing them to a higher risk of musculoskeletal injuries [36].

In extreme dehydration in a hot temperature condition, cardiovascular impairment will result in a complete lack of sweating (anhidrosis), and consequently, thermoregulation is threatened [24, 37], clinically manifested by hyperthermia, hyperventilation, and cardiovascular stress reducing the brain, musculoskeletal, and skin blood perfusion, causing muscle and cellular glycogen utilization and neural activity disorders. In more severe cases, it progresses to heat cramps, heat exhaustion, heat collapse, heatstroke, brain swelling, kidney failure, coma, or even death [38, 39].

All these points strengthen the importance of hydration for sports practice, preventing cardiovascular, neurological, and physical distress and enhancing the athletes' performance. Management of proper hydration remains challenging in sports, as athletes chiefly neglect it in their diet [35]. Therefore, physicians and coaches must be aware of reinforcing the importance of euhydration and making athletes conscious, moreover, paying close attention to symptoms and clinical signs commonly related to dehydration, such as headache, dark yellow- or amber-colored urine, decreased urination, dry mouth and mucous membrane, low body temperature, muscle cramps, vomiting, diarrhea, and an inability of to drink fluids [8, 40].

Thus, educating athletes about voluntary drinking is part of the strategy, but individualizing hydration needs should be mandatory, taking into account the athlete's sweat rates. Moreover, in hot environments, the intensity of training or competition, clothing, acclimatization (the modifiable factors), and ambient conditions should be carefully considered to achieve a proper hydration strategy to guarantee athletes' performance [41, 42].

2.4.4.4 Athlete's Overdrinking Due to Excessive Concern About Hydration

In some situations, a huge concern about keeping the body well hydrated could leave athletes to overdrink fluids in excess and, if associated with a more significant loss of electrolytes through sweat and urine, can result in hyponatremia (plasma sodium <135 mmol/L), so-called water intoxication. Clinically, the symptoms may vary from nausea, vomiting, confusion or delirium, seizures, respiratory distress, loss of consciousness, and even death if not treated adequately.

Therefore, an adequate strategy for hydration should consider the expected water and electrolyte loss according to the duration and intensity of physical demands during training or competition, the environmental conditions, and the athlete's acclimatization status [34, 37].

2.4.4.5 The Concern About Athlete's Hydration During Competition

Athlete's hydration during competition has received close attention from sports organizations, mainly in tropical countries such as Brazil, where during summer hot temperatures are often high, therefore, exposing athletes to dehydration and consequently predisposing them to heat stroke. In this context, the Brazilian Football Confederation (Soccer) (CBF) has provided the referee to interrupt the game if the environment temperature exceeds 28 °C or at his discretion, allowing the athletes to hydrate for twenty minutes. FIFA Executive Committee has also approved the regulation of technical stoppages in soccer games with temperatures above 31 °C [43–45].

2.5 Unique Heat-Related Illness in Sports and Exercise

Clinically, it can vary from edema and heat rash, muscle cramps and spasms, heat syncope, exertional-related heat exhaustion, exertional rhabdomyolysis, and heat stroke—a life-threatening condition (Table 2.1) [46]. Heat edema and heat rash are the mildest clinical presentations of heat-related illness.

2.5.1 Heat Edema

Heat edema generally manifests by swelling of soft tissues in the lower limbs, secondary to peripheral vasodilatory pooling. It drives the blood flow to the skin surface to dissipate the heat by sweat and direct radiation, increasing the blood flow to the legs and hands by gravity and

Table 2.1 Heat-related illness in sports and exercise

Exercise heat-related illness	Body core temperature	Symptoms and clinical findings
Heat edema	Normal	No trauma history Mild and subtle enlargement of the hands and feet is commonly present, signing to heat-related issues (hands, ankles, and feet) (secondary to peripheral vasodilatory pooling, increasing the blood flow to the legs and hands by following gravity)
Heat rash	Normal	Red clusters of small pimples or blisters Where clothing occludes and causes friction (On the neck, upper chest, groin, under the breast, and elbows)
Heat cramps	Normal	Painful skeletal muscle cramps or spasms High-demanding muscle groups (quadriceps, hamstrings, gastrocnemius, arms, trunk, and abdominal muscles)
Heat syncope	**37 °C–40 °C**	Dizziness, weakness, fainting.
Heat exhaustion	**37 °C–40 °C**	Fatigue, nausea, vomiting, headache, dizziness, irritability, heavy sweating, tachycardia, hypotension, fast breathing, mild and temporary confusion, extreme weakness
Rhabdomyolysis (muscle breakdown)	**> 39 °C**	Muscle weakness and pain Dark urine or reduced urine input
Heat stroke	**> 40 °C**	Confusion, emotional instability, visual disturbances, seizures, mental confusion, and loss of consciousness Tachycardia and hypotension Possible history of heat exhaustion symptoms before mental change

aiming to maintain and balance the core temperature [46].

It increases the permeability and hydrostatic intravascular pressure, moving the fluid to the interstitium. The inability of the lymph system to clear and balance the excess fluid causes an accumulation of fluid in the surrounding tissues, the pooling fluid—*the edema.*

In general, there is no report of trauma history, and clinical findings of mild and subtle enlargement of the hands and feet are commonly present, signing to heat-related issues. Moreover, the sports physician must be aware of some differential diagnoses because they could also present potential causes of edema, such as underlying cardiopathy, diabetes mellitus, and liver or kidney diseases or failures [47].

The edema management involves gently elevating the affected limbs and encouraging relative rest. Compression stockings may be recommended in cases where resolution is slow. Ensuring the athlete maintains proper hydration and salt intake is crucial, as these factors can influence recovery duration. Diuretics should be avoided, as they may worsen the condition by further reducing intravascular volume. Generally, improvement is anticipated within one or two weeks with acclimatization or sooner upon the athlete's return to their accustomed environment [29].

2.5.2 Heat Rash

Heat rash, prickly heat, or miliaria rubra is manifested clinically by red clusters of small pimples or blisters, mainly found where clothing occludes and causes friction, such as on the neck, upper chest, groin, under the breast, and elbows. When exposed to high heat and humidity, which lead to excessive sweating, individuals can develop heat rash, characterized by a rapid-onset pruritic papulovesicular rash. The rash may have a "prickling" sensation [29, 48].

The pathophysiology involves obstructing sweat glands, allowing eccrine sweat to leak into the epidermis or dermis. Measures to cool the affected area, keep the area dry, apply powder, not use ointments and cream, and minimize clothing coverage can facilitate resolution.

It commonly requires a week or more to resolve fully. Application of a mild anti-inflammatory lotion, such as desonide, may offer symptom relief and accelerate the healing process. The differential diagnosis for such a rash may include viral exanthem, allergic contact dermatitis, or urticaria [48].

2.5.3 Skeletal Muscle Cramps and Spasms

Although there is no evidence of a close relationship between high temperature and the occurrence of skeletal muscle cramps or spasms, they are one of the initial manifestations of heat-related illness. Moreover, it may occur independently or concomitant with heat exhaustion manifestations. There is a scenario of excessive perspiration, lengthy exposition of high temperatures, and insufficient intake of water and electrolytes that causes spasmodic muscle contraction and discomfort [10, 47].

Sodium depletion is widely known to exacerbate heat-induced cramps and spams; however, the role of magnesium and calcium imbalances in heat-related cramp genesis remains inconclusive. Dehydration, sodium and/or potassium imbalance, exposure to severe environmental conditions, and neurogenic fatigue are believed factors implicated in heat-related cramps or spams [29, 47].

Painful cramps generally touch high-demanding muscle groups submitted to exercises longer than two hours, such as the quadriceps, hamstrings, gastrocnemius, and abdominal muscles.[6] As it presents a benign and self-limited condition, gentle stretching, ice cooling application, massage, and temporary stopping of physical activity yield positive outcomes. Moreover, close attention to oral fluid and electrolyte repletion is critical, although its efficacy is limited in acute cases [29].

2.5.4 Heat Syncope

It can occur during excessive heat exposure, where the blood flow is shuttled to the skin and limb extremities. In this condition, rapid and sudden changes from sitting to standing or bending posture may result in a temporary drop in blood pressure that leads the individual to faint. Recovery tends to be quick once the athlete is laid flat, as this positioning normalizes blood flow. However, in cases of a rapid loss of consciousness, it could lead to a concussion, and therefore, the athlete should be carefully assessed before returning to training or competition [25, 29, 48].

2.5.5 Exercise-Related Heat Exhaustion

Exercise-related heat exhaustion is commonly seen in athletes and military recruits in basic training [10]. It results from excessive heat during exercise or sports practice, which causes the body's temperature to rise above normal. This condition occurs when the body cannot dissipate the extra heat produced during exercise, increasing the core temperature higher than is healthy. It can be deleteriously potentialized when concomitant dehydration conditions take place. Another setting that may predispose to exercise-related heat exhaustion is an environment with high humidity, as it impairs the body's sweating and compromises one of the strategies of the body's cooling [10, 23, 24].

Clinical findings and symptoms are fatigue, nausea, vomiting, headache, dizziness, irritability, heavy sweating, tachycardia, hypotension, fast breathing, mild and temporary confusion, and extreme weakness. Although heat exhaustion is less severe than heat stroke, it can evolve into it; therefore, the physician must be aware [49].

In severe cases or recurrence of symptoms, volumetric repositing of NaCl at 0.95 should be taken into account as it could allow a prompt resolution of symptoms. An important point to be emphasized is that the color of the urine after training or competition is an efficient clinical parameter to assess the athlete's hydration pattern, where the darker color of urine (copper brown until ruby brown) sign of dehydration and the need for more hydration [50].

2.5.6 Exertional Rhabdomyolysis

Exertional rhabdomyolysis is a clinical condition caused by the breakdown of muscle-skeletal cells secondary to heavy endurance events in heat stress conditions. Moreover, muscle sports trauma or crushing injuries, toxins, infections, and medication may also present additional causes [51].

The pathophysiology involves muscle tissue breakdown and necrosis, releasing electrolytes and large proteins into the bloodstream and increasing serum creatine kinase and myoglobin levels. The main clinical findings are muscle cramps, pain, exercise intolerance, and weakness. Although reddish-brown urine of myoglobinuria is another clinical finding related to this condition, it could be absent due to the rapid myoglobin clearance. Heartbeat arrhythmia, seizures, and kidney overloading follow it [52].

The diagnosis is established by clinical history, anamnesis, and physical exam, and the laboratory findings show an acute elevation of the serum creatine kinase (CK) and other muscle enzymes; usually, the serum CK levels are at least five times the upper limit of normal. The treatment focuses on rest, clinical support with IV fluids, and approaching the inciting causes and resulting complications [51, 52].

2.5.7 Exertional Heat Stroke

In exertional heat stroke (EHT), the body temperature rises to 40 °C or higher within 10–15 min following a collapse during intense physical activity and excessive sun exposure or intense heat. In this condition, the body thermoregulation mechanism is inefficient in reducing and controlling the rapid increase of body temperature, and the sweating mechanism fails [1, 25, 53].

EHT generally results from an association of factors, such as the athlete's excessive motivation ("warrior mentality"), exhaustive training and dehydration, heat and humidity, and lack of acclimatization. Moreover, the athletes' clothing and protective gear could also reduce the heat change transfer to the environment, as seen in sports such as American football, car racing, etc. EHT is also seen in the military and highly active individuals as they push their body limits and are also exposed to heat adverse conditions, predisposing them to EHT [7, 10, 54].

Using stimulants and the addiction to amphetamine and cocaine could be potential causes, as they accelerate metabolism and increase the body's temperature. Other drugs such as antihistaminic, antispasmodic, and antidepressant could also reduce transpiration. Therefore, sports physicians should be aware of this.

EHT results in central nervous dysfunction (CNS) that could be manifested through a diverse array of symptoms and signs, including disorientation, emotional instability, visual disturbances, seizures, mental confusion, and loss of consciousness. Most patients typically exhibit tachycardia and hypotension. Additional symptoms observed are headache, dizziness, hot, dry skin or profuse sweating, high body temperature, hyperventilation, nausea, vomiting, diarrhea, weakness, thirst, muscle cramps, a deficit of muscle function, and ataxia [1, 54, 55].

The sports physician should be aware that the prognosis worsens if the core temperature persists above the critical threshold of 40.5–41.0 °C. It could be fatal if not treated promptly and adequately. The morbidity and mortality of EHS arise directly from ischemia and oxidative stress (oxygen metabolic disorder). Some studies have confirmed these observations, reporting a close relationship between the duration of high body temperature and morbidity/mortality. A series of cases of heat stroke reports survival tends to increase when the core temperature is cooled below 38.9 °C within 60 min. Another study pointed out higher survival rates even when cooling to the same threshold was achieved within 30 min [55, 56].

2.6 Management and Extraction of Heat-Related Illness

Some steps to be considered when managing and rescuing an athlete suffering from a heat stroke.

2.6.1 When Cooling Must be Started?

As this is a life-threatening condition, the sports physician must remember that the principle is "Cool first, transport second." Moreover, the heat stroke survival rate is strictly related to cooling starting within the first 10 min of collapse. Therefore, *cooling must begin in the field immediately after a brief and fast assessment, and as it is the golden half-hour, it* must not be delayed. The equipment and clothes must be removed simultaneously after the cooling starts [47, 57].

2.6.2 Which Methods Can be Applied to Achieve an Athlete's Body Cooling?

Conductive cooling in ice/water cold immersion (CWI) is the most effective strategy for cooling the athlete's body; however, a tub cannot be available or feasible everywhere, so this cooling intervention cannot be accessible for most. However, alternative cooling methods can also be applied, following a descending order from the highest cooling rates, such as cold water immersion in a tarp or body bag cooling, ice packs/chilled bed, ice sheet and towel rotation, and mist water and fan [57–59].

2.6.3 How Can the Athlete's Core Temperature Response to Cooling Be Monitored?

Cooling monitoring is based on assessing the core body temperature, mental status, and vital signs (air, breath, circulation) every 10 minutes. Rectal thermometry is the gold standard for evaluating and monitoring the core body temperature. Cooling measures should be repeated if the athlete's core temperature rises until it reaches and is kept at under 39 °C. On the other hand, if the core temperature drops toward 36 °C in the 15 minutes post-cooling, the athlete should be dried, warped into blankets to be warmed, and transferred to a hospital for advanced medical care. Exposing the athlete to sunlight can also be a valuable warming strategy while awaiting a hospital transfer [60–62].

2.6.4 When Can the Athlete Be Transferred to a Hospital?

Transferring to a hospital can be done *when adequate cooling has been verified and confirmed* by an athlete's body temperature assessment, a core temperature of 39 °C or below, or *if the cooling measures are inefficient.*

2.6.5 Management of Heat Stroke

In clinical practice, once the diagnosis of EHS is clinically confirmed, measures should be implemented to reduce body temperature promptly. Although rapid cooling through cold water immersion is the foremost approach to mitigating exertional heat stroke (EHS), it requires an infrastructure that is not available in some instances.

Start cooling and simultaneously remove or swiftly discard clothing or equipment that could make the body cooling process difficult. Submit the athlete to a cold shower or administer cold water using a hose [11, 52, 53]. Place the athlete in a cool, shaded area to apply cold water or ice-damp towels to maximize cooling. Moreover, as standard practice, a towel replacement should be done upon losing freshness or every two or three minutes. Another possibility is to wrap the athlete with a tarp or sheet with a large amount of ice and follow a continuous ice replacement when melting is observed.

EHS does not entail a deviation in the hypothalamic set point akin to febrile conditions, so there are no specific pharmacological interventions to approach it. However, it is noteworthy that there is growing interest among researchers regarding the potential correlation between malignant hyperthermia (MH) and exertional heat stroke. Even though dantrolene is under scrutiny as a potential adjunctive therapy for EHS, because it is conventionally used in cases of malignant hyperthermia, further investigations are needed to support their clinical application. Similarly, the use of immune modulators seems plausible in treating HST, presenting a field of ongoing research as a therapeutic avenue for severe exertional heat illness; a comprehensive analysis to confirm their efficacy and safety is also needed [54–56].

2.7 Prevention

Preventing is always the best strategy for treating a life-threatening condition like heat stroke. Besides, some measures can be implemented to minimize the occurrence of this condition and its harmful effects.

2.7.1 Do Misting Stations Work?

Misting stations in sports could be an alternative to providing humidification and cooling sweltering heat spaces or environments, especially during the summer. They were adopted in the Tokyo 2020 Summer Olympics for athletes and horses to cool them and promote relief from the high temperatures of almost 35 °C or more. Therefore, misting systems could help minimize the harmful effects of high temperatures and maintain ideal temperatures during sports training or competition [63, 64].

2.7.2 When Should the Competition Be Stopped Due to High Temperatures?

Wet Bulb Globe Temperature (WGBT) is a device that measures heat, humidity, and other factors to provide a more reliable feature of environmental conditions. FIFA's position at extreme heat is that in a WGBT above 32 °C (89.6 °F) scenario, cooling breaks should be implemented in both halves of the game, and the decision to suspend or cancel the game is analyzed by the referee and competition organizers' discretion. After recurrent episodes of heat-related illness affecting ballboys, attendants, and players, in 2019, the Australian Open Tennis Tournament introduced the "AO Heat Stress Scale." This scale starts with the temperature conditions (radiant heat, air temperature in the shade, relative humidity, and wind speed) and goes up to the game's cancellation. The US Soccer Federation Guidelines consider that 40.56 °C (105 °F) and with a heat index dangerous level is a condition to cancel or postpone training or games.

Besides these recommendations, athletes must remember that hydration must be done before, during, and after sports practice. Light clothing potentializes the body's heat change to the environment. Rest breaks and active cooling benefit the body's thermoregulation. Respecting and paying close attention to your body's limits and stopping exercising or sports practice when the body requests is the best way to keep healthy and safe [65–68].

2.7.3 Do Cooling Jacket Systems Prevent the Incidence of Injuries?

Cooling jacket systems involve technologies to regulate body temperature, offering a cooling relief from hot environments. In this context, *a fan-cooling system and/or cooling jacket could mitigate thermal strain and discomfort from exercising in hot, humid environments.* Consequently, it *could reduce the risks of heat-related illness, dehydration, and a heat stroke.* However, even though using cooling jackets in sports could indirectly maintain a safe exercise environment and benefit perceptual response to heat and thermos-physiological behavior and sports performance, further studies are needed to confirm its application [69, 70].

2.7.4 Does Acclimatization Prevent Injuries?

Acclimatization is, without a doubt, a valuable strategy to minimize harmful effects from adverse environmental conditions, such as one with high temperatures. Physiologically, acclimatization triggers adaptations that enhance sweat and blood flow from the skin, reduce body temperature and cardiovascular stress, promote fluid balance, alter metabolism, and provide excellent cellular protection. Therefore, an athlete's acclimatization and conditioning to a more intense heat environment improve their thermoregulation and help them face it. Consequently, athlete's acclimatization and conditioning to intense heat environments are vital measures in preventing heat illness, and sports physicians and trainers should be aware of that [68, 71–74].

2.7.5 Preventing Measures for Heat-Related Illness in Sports

- Take fluids during sports training.
- Hydration is critical for the body's thermoregulation.
- The strategy of hydration should consider the expected water and electrolyte loss considering the duration and intensity of physical demands during training or competition.
- The environmental conditions should be carefully assessed before training or competing.
- The athlete's acclimatization may prevent or even reduce the occurrence of heat-related illness.
- Wear proper clothing to sports practice.
- Pay close attention to the weather conditions.

2.8 Take-Home Message on Heat-Related Illness in Sports

- Exercising in hot and humid environment overloads the body's thermoregulation.
- Dehydration and a hot environment harm sports performance.
- Heat edema and heat rash are the mildest clinical presentations.
- Heat-related exhaustion is less serious than heat stroke, but it can progress to heat stroke if treatment is inadequate or delayed.
- Overdrinking fluids and significant loss of electrolytes through sweat and urine can result in hyponatremia.
- Reddish-brown urine of myoglobinuria may be a clinical sign of exertional rhabdomyolysis.
- Laboratory exams should be performed to investigate and rule out differential diagnosis.
- Using stimulants and the addiction to amphetamine and cocaine accelerate metabolism and increase the body's temperature; therefore, social habits should be investigated.

Acknowledgments We wish to thank Ana Karina Piedade and Mariana Percario Piedade for their help in developing and producing the drawings, storyboard, schema, and figures for this chapter.

References

1. Yeargin SW, Dickinson JJ, Emerson DM, Koller J, Torres-McGehee TM, Kerr ZY. Exertional heat illness risk factors and physiological responses of youth football players. J Sport Health Sci. 2021;10:91–8. https://doi.org/10.1016/j.jshs.2019.03.002.
2. Cooper ER, Ferrara MS, Casa DJ, Powell JW, Broglio SP, Resch JE, et al. Exertional heat illness in American football players: when is the risk greatest? J Athl Train. 2016;51:593–600. https://doi.org/10.4085/1062-6050-51.8.08.
3. Maughan RJ, Watson P, Shirreffs SM. Heat and cold. Sports Med. 2007;37:396–9. https://doi.org/10.2165/00007256-200737040-00032.

4. Marc A, Sedeaud A, Guillaume M, Rizk M, Schipman J, Antero-Jacquemin J, et al. Marathon progress: demography, morphology and environment. J Sports Sci. 2014;32:524–32. https://doi.org/10.1080/02640414.2013.835436.
5. Manning C, Benjamin C, Sekiguchi Y, Butler C, Szymanski M, Stearns R, et al. Environmental stress symptoms during heat acclimatization, heat acclimation, and intermittent heat training. Int J Environ Res Public Health. 2023;20:3219. https://doi.org/10.3390/ijerph20043219.
6. Lorenzo S, Halliwill JR, Sawka MN, Minson CT. Heat acclimation improves exercise performance. J Appl Physiol. 2010;109:1140–7. https://doi.org/10.1152/japplphysiol.00495.2010.
7. Tripp BL, Winkelmann ZK, Eberman LE, Smith MS. Factors affecting incidence rate of exertional heat illnesses: analysis of 6 years of high school football practices in north Central Florida. Orthop J Sports Med. 2021;9:232596712110266. https://doi.org/10.1177/23259671211026627.
8. Maron BJ, Doerer JJ, Haas TS, Tierney DM, Mueller FO. Sudden deaths in young competitive athletes. Circulation. 2009;119:1085–92. https://doi.org/10.1161/CIRCULATIONAHA.108.804617.
9. Bouchama A, Abuyassin B, Lehe C, Laitano O, Jay O, O'Connor FG, et al. Classic and exertional heatstroke. Nat Rev Dis Primers. 2022;8:8. https://doi.org/10.1038/s41572-021-00334-6.
10. Périard JD, Eijsvogels TMH, Daanen HAM. Exercise under heat stress: thermoregulation, hydration, performance implications, and mitigation strategies. Physiol Rev. 2021;101:1873–979. https://doi.org/10.1152/physrev.00038.2020.
11. Gamage PJ, Fortington LV, Finch CF. Epidemiology of exertional heat illnesses in organised sports: a systematic review. J Sci Med Sport. 2020;23:701–9. https://doi.org/10.1016/j.jsams.2020.02.008.
12. Yeargin SW, Dompier TP, Casa DJ, Hirschhorn RM, Kerr ZY. Epidemiology of exertional heat illnesses in National Collegiate Athletic Association Athletes during the 2009–2010 through 2014–2015 academic years. J Athl Train. 2019;54:55–63. https://doi.org/10.4085/1062-6050-504-17.
13. Inoue H, Tanaka H, Sakanashi S, Kinoshi T, Numata H, Yokota H, et al. Incidence and factor analysis for the heat-related illness on the Tokyo 2020 Olympic and Paralympic games. BMJ Open Sport Exerc Med. 2023;9:e001467. https://doi.org/10.1136/bmjsem-2022-001467.
14. Cramer MN, Gagnon D, Laitano O, Crandall CG. Human temperature regulation under heat stress in health, disease, and injury. Physiol Rev. 2022;102:1907–89. https://doi.org/10.1152/physrev.00047.2021.
15. Geneva II, Cuzzo B, Fazili T, Javaid W. Normal body temperature: a systematic review. Open Forum Infect Dis. 2019;6:ofz032. https://doi.org/10.1093/ofid/ofz032.
16. Lim CL, Byrne C, Lee JK. Human thermoregulation and measurement of body temperature in exercise and clinical settings. Ann Acad Med Singap. 2008;37:347–53.
17. Obermeyer Z, Samra JK, Mullainathan S. Individual differences in normal body temperature: longitudinal big data analysis of patient records. BMJ. 2017;359:j5468. https://doi.org/10.1136/bmj.j5468.
18. Haman F, Souza SCS, Castellani JW, Dupuis M-P, Friedl KE, Sullivan-Kwantes W, et al. Human vulnerability and variability in the cold: establishing individual risks for cold weather injuries. Temperature. 2022;9:158–95. https://doi.org/10.1080/23328940.2022.2044740.
19. Jay O, Gariépy LM, Reardon FD, Webb P, Ducharme MB, Ramsay T, et al. A three-compartment thermometry model for the improved estimation of changes in body heat content. Am J Phys Regul Integr Comp Phys. 2007;292:R167–75. https://doi.org/10.1152/ajpregu.00338.2006.
20. Romanovsky AA. Thermoregulation: some concepts have changed. Functional architecture of the thermoregulatory system. Am J Physiol Regul Integr Comp Physiol. 2007;292:R37–46. https://doi.org/10.1152/ajpregu.00668.2006.
21. Sawka MN, Cheuvront SN, Kenefick RW. High skin temperature and hypohydration impair aerobic performance. Exp Physiol. 2012;97:327–32. https://doi.org/10.1113/expphysiol.2011.061002.
22. Blatteis CM. Physiology and pathophysiology of temperature regulation. Thermal physiology: brief history and perspectives. World Scientific; 1998. p. 3–12.
23. Montain SJ, Ely MR, Cheuvront SN. Marathon performance in thermally stressing conditions. Sports Med. 2007;37:320–3. https://doi.org/10.2165/00007256-200737040-00012.
24. Cheshire WP. Thermoregulatory disorders and illness related to heat and cold stress. Auton Neurosci. 2016;196:91–104. https://doi.org/10.1016/j.autneu.2016.01.001.
25. Westwood CS, Fallowfield JL, Delves SK, Nunns M, Ogden HB, Layden JD. Individual risk factors associated with exertional heat illness: a systematic review. Exp Physiol. 2021;106:191–9. https://doi.org/10.1113/EP088458.
26. Ebi KL, Capon A, Berry P, Broderick C, de Dear R, Havenith G, et al. Hot weather and heat extremes: health risks. Lancet. 2021;398:698–708. https://doi.org/10.1016/S0140-6736(21)01208-3.
27. Madden CJ, Morrison SF. Central nervous system circuits that control body temperature. Neurosci Lett. 2019;696:225–32. https://doi.org/10.1016/j.neulet.2018.11.027.
28. Gagge AP, Gonzalez RR. Mechanisms of heat exchange: biophysics and physiology. Compr Physiol. 1996:45–84. https://doi.org/10.1002/cphy.cp040104.
29. Nichols AW. Heat-related illness in sports and exercise. Curr Rev Musculoskelet Med. 2014;7:355–65. https://doi.org/10.1007/s12178-014-9240-0.

30. Cramer MN, Jay O. Biophysical aspects of human thermoregulation during heat stress. Auton Neurosci. 2016;196:3–13. https://doi.org/10.1016/j.autneu.2016.03.001.
31. Bernard P, Chevance G, Kingsbury C, Baillot A, Romain A-J, Molinier V, et al. Climate change, physical activity and sport: a systematic review. Sports Med. 2021;51:1041–59. https://doi.org/10.1007/s40279-021-01439-4.
32. Maughan RJ, Shirreffs SM. Dehydration and rehydration in competative sport. Scand J Med Sci Sports. 2010;20:40–7. https://doi.org/10.1111/j.1600-0838.2010.01207.x.
33. Jacobs PJ, Oosthuizen MK, Mitchell C, Blount JD, Bennett NC. Heat and dehydration induced oxidative damage and antioxidant defenses following incubator heat stress and a simulated heat wave in wild caught four-striped field mice Rhabdomys dilectus. PLoS One. 2020;15:e0242279. https://doi.org/10.1371/journal.pone.0242279.
34. Trangmar SJ, González-Alonso J. Heat, Hydration and the human brain, heart and skeletal muscles. Sports Med. 2019;49:69–85. https://doi.org/10.1007/s40279-018-1033-y.
35. Bergeron MF. Hydration and thermal strain during tennis in the heat. Br J Sports Med. 2014;48:i12–7. https://doi.org/10.1136/bjsports-2013-093256.
36. Judge L, Bellar D, Popp J, Craig B, Schoeff M, Hoover D, et al. Hydration to maximize performance and recovery: knowledge, attitudes, and behaviors among collegiate track and field throwers. J Hum Kinet. 2021;79:111–22. https://doi.org/10.2478/hukin-2021-0065.
37. Desai Y, Khraishah H, Alahmad B. Heat and the heart. Yale J Biol Med. 2023;96:197–203. https://doi.org/10.59249/HGAL4894.
38. Filep EM, Murata Y, Endres BD, Kim G, Stearns RL, Casa DJ. Exertional heat stroke, modality cooling rate, and survival outcomes: a systematic review. Medicina (B Aires). 2020;56:589. https://doi.org/10.3390/medicina56110589.
39. McCubbin AJ, Allanson BA, Caldwell Odgers JN, Cort MM, Costa RJS, Cox GR, et al. Sports dietitians Australia position statement: nutrition for exercise in hot environments. Int J Sport Nutr Exerc Metab. 2020;30:83–98. https://doi.org/10.1123/ijsnem.2019-0300.
40. McCartney D, Desbrow B, Irwin C. The effect of fluid intake following dehydration on subsequent athletic and cognitive performance: a systematic review and meta-analysis. Sport Med Open. 2017;3:13. https://doi.org/10.1186/s40798-017-0079-y.
41. Mosler S, Braun H, Carlsohn A, Großhauser M, König D, Lampen A, et al. Position of the working group sports nutrition of the German nutrition society (DGE): fluid replacement in sports. Deutsche Zeitschrift Für Sportmedizin/Ger J Sport Med. 2020;71:178–84. https://doi.org/10.5960/dzsm.2020.453.
42. Baker LB. Sweating rate and sweat sodium concentration in athletes: a review of methodology and intra/Interindividual variability. Sports Med. 2017;47:111–28. https://doi.org/10.1007/s40279-017-0691-5.
43. Magro DO. Nutrition and hydration. In: The sports medicine physician. Cham: Springer International Publishing; 2019. p. 85–98. https://doi.org/10.1007/978-3-030-10433-7_8.
44. Da Silva AI, Fernandez R. Dehydration of football referees during a match. Br J Sports Med. 2003;37:502–6. https://doi.org/10.1136/bjsm.37.6.502.
45. Houssein M, Lopes P, Fagnoni B, Ahmaidi S, Yonis SM, Leprêtre P-M. Hydration: the new FIFA world Cup's challenge for referee decision making? J Athl Train. 2016;51:264–6. https://doi.org/10.4085/1062-6050-51.3.04.
46. Howe AS, Boden BP. Heat-related illness in athletes. Am J Sports Med. 2007;35:1384–95. https://doi.org/10.1177/0363546507305013.
47. Gauer R, Meyers BK. Heat-related illnesses. Am Fam Physician. 2019;99:482–9.
48. MacKnight JM, Mistry DJ. Allergic disorders in the athlete. Clin Sports Med. 2005;24:507–23. https://doi.org/10.1016/j.csm.2005.04.003.
49. Danzig RM, Raunig JM, Acholonu CJ. Exertional heat illness—from identifying heat rash to treating heat stroke. Pediatr Ann. 2024;53:53. https://doi.org/10.3928/19382359-20231113-04.
50. Roberts WO, Armstrong LE, Sawka MN, Yeargin SW, Heled Y, O'Connor FG. ACSM expert consensus statement on exertional heat illness: recognition, management, and return to activity. Curr Sports Med Rep. 2023;22:134–49. https://doi.org/10.1249/JSR.0000000000001058.
51. He S, Li R, Peng Y, Wang Z, Huang J, Meng H, et al. ACSL4 contributes to ferroptosis-mediated rhabdomyolysis in exertional heat stroke. J Cachexia Sarcopenia Muscle. 2022;13:1717–30. https://doi.org/10.1002/jcsm.12953.
52. Fernandes PM, Davenport RJ. How to do it: investigate exertional rhabdomyolysis (or not). Pract Neurol. 2019;19:43–8. https://doi.org/10.1136/practneurol-2018-002008.
53. Asmara IGY. Diagnosis and management of heatstroke. Acta Med Indones. 2020;52:90–7.
54. Garcia CK, Renteria LI, Leite-Santos G, Leon LR, Laitano O. Exertional heat stroke: pathophysiology and risk factors. BMJ Med. 2022;1:e000239. https://doi.org/10.1136/bmjmed-2022-000239.
55. Knapik JJ, Epstein Y. Exertional heat stroke: pathophysiology, epidemiology, diagnosis, treatment, and prevention. J Spec Oper Med. 2019;19:108. https://doi.org/10.55460/5P2Q-1MBQ.
56. Glazer JL. Management of heatstroke and heat exhaustion. Am Fam Physician. 2005;71:2133–40.
57. Douma MJ, Aves T, Allan KS, Bendall JC, Berry DC, Chang W-T, et al. First aid cooling techniques

for heat stroke and exertional hyperthermia: a systematic review and meta-analysis. Resuscitation. 2020;148:173–90. https://doi.org/10.1016/j.resuscitation.2020.01.007.
58. Brade C, Dawson B, Wallman K, Polglaze T. Postexercise cooling rates in 2 cooling jackets. J Athl Train. 2010;45:164–9. https://doi.org/10.4085/1062-6050-45.2.164.
59. Naito T, Saito T, Ohhashi M, Hayashi S. Recovery with a fan-cooling jacket after exposure to high solar radiation during exercise in hot outdoor environments. Front Sports Act Living. 2023;5:1106882. https://doi.org/10.3389/fspor.2023.1106882.
60. National Centre for Disease Control Government of India Acknowledgement. Emergency cooling for severe heat related illnesses. n.d.
61. Bouchama A, Dehbi M, Chaves-Carballo E. Cooling and hemodynamic management in heatstroke: practical recommendations. Crit Care. 2007;11:R54. https://doi.org/10.1186/cc5910.
62. Gaudio FG, Grissom CK. Cooling methods in heat stroke. J Emerg Med. 2016;50:607–16. https://doi.org/10.1016/j.jemermed.2015.09.014.
63. Honda Moore S. How to cool a city struggling with heatwaves. Copyright 2024 BBC. n.d.
64. France Médias Monde – Nanterre Trade and Companies Register: 501 524 029. Mist and morning glories: Tokyo 2020 bids to beat the heat. 92130 Issy Les Moulineaux – France. n.d.
65. Gouttebarge V, Duffield R, den Hollander S, Maughan R. Protective guidelines and mitigation strategies for hot conditions in professional football: starting 11 hot tips for consideration. BMJ Open Sport Exerc Med. 2023;9:e001608. https://doi.org/10.1136/bmjsem-2023-001608.
66. Jay O, Broderick C, Smallcombe J. Extreme heat policy – the new sports medicine Australia (SMA) extreme heat Policy. n.d.
67. Roberts WO, Armstrong LE, Sawka MN, Yeargin SW, Heled Y, O'Connor FG. ACSM expert consensus statement on exertional heat illness: recognition, management, and return to activity. Curr Sports Med Rep. 2021;20:470–84. https://doi.org/10.1249/JSR.0000000000000878.
68. Périard JD, Racinais S, Sawka MN. Adaptations and mechanisms of human heat acclimation: applications for competitive athletes and sports. Scand J Med Sci Sports. 2015;25(Suppl 1):20–38. https://doi.org/10.1111/sms.12408.
69. Fernández-Lázaro D, García JF, Corchete LA, Del Valle Soto M, Santamaría G, Seco-Calvo J. Is the cooling vest an ergogenic tool for physically active individuals? Assessment of perceptual response, Thermo-physiological behavior, and sports performance: a systematic review and meta-analysis. Bioengineering (Basel). 2023;10:132. https://doi.org/10.3390/bioengineering10020132.
70. Duffield R, Dawson B, Bishop D, Fitzsimons M, Lawrence S. Effect of wearing an ice cooling jacket on repeat sprint performance in warm/humid conditions. Br J Sports Med. 2003;37:164–9. https://doi.org/10.1136/bjsm.37.2.164.
71. Racinais S, Alonso JM, Coutts AJ, Flouris AD, Girard O, González-Alonso J, et al. Consensus recommendations on training and competing in the heat. Br J Sports Med. 2015;49:1164–73. https://doi.org/10.1136/bjsports-2015-094915.
72. Gibson OR, James CA, Mee JA, Willmott AGB, Turner G, Hayes M, et al. Heat alleviation strategies for athletic performance: a review and practitioner guidelines. Temperature (Austin). 2020;7:3–36. https://doi.org/10.1080/23328940.2019.1666624.
73. Waddington GS. Performance in the heat: health complaints and heat stress prevention strategies as predictors. J Sci Med Sport. 2020;23:321. https://doi.org/10.1016/j.jsams.2020.02.001.
74. Tyler CJ, Reeve T, Hodges GJ, Cheung SS. The effects of heat adaptation on physiology, perception and exercise performance in the heat: a meta-analysis. Sports Med. 2016;46:1699–724. https://doi.org/10.1007/s40279-016-0538-5.

It's Cold

3

Chantal Nguyen, Seth L. Sherman,
and Michael Fredericson

3.1 Introduction to Specific Environment and Environmental Challenges

Human adaptation to exposure and survival in the cold has evolved greatly over the years. In early civilization, the demographic most affected by cold-induced injuries involved outdoor laborers, such as fisherman or hunters, whose livelihoods were linked to trades requiring prolonged exposure to cold, wet, or high-altitude environments [1]. As industrial avenues for cold storage improved, those most affected by cold were meat/storage employees who worked in refrigerated environments or the homeless population without access to shelter [2, 3]. High rates of injury and fatality from prolonged cold exposure occurred since the human body failed to maintain temperature homeostasis for basal metabolic functioning [4].

C. Nguyen (✉) · M. Fredericson
Department of Orthopedic Surgery, Division of Physical Medicine & Rehabilitation, Stanford University, Stanford, CA, USA
e-mail: nguyenc3@stanford.edu; mfred2@stanford.edu

S. L. Sherman
Department of Orthopedic Surgery, Division of Physical Medicine & Rehabilitation, Stanford University, Stanford, CA, USA

Department of Orthopedic Surgery, Stanford University, Stanford, USA
e-mail: shermans@stanford.edu

Prolonged exposure to cold negatively affects the natural temperature equilibrium within the body, which is the ideal temperature range (37 degrees Celsius) required for biochemical processes, namely, enzymatic function, to maintain a basic metabolic rate for survival [4, 5]. Cold exposure induces peripheral vasoconstriction to prevent loss of heat from the periphery via the natural process of convection. To compensate for cold environments, the body undergoes thermogenesis, thus increasing involuntary behaviors such as shivering to yield heat [4, 5]. Shivering amplifies resting metabolic rate by up to five to six times that of baseline values, but at the expense of increased energy expenditure, which, over time, promotes exhaustion and fatigue [4–7]. Eventually, if the body attempts thermogenesis for too long, glycogen energy stores are depleted, which impairs organ function. Increased rates of cold-induced injury, morbidity, and mortality result [4, 8].

These effects of prolonged cold exposure and aversion to extreme cold initially led humans to migrate to warmer places for survival or forced adaptation to temperature extremes [9]. These adaptations included access to warmer clothes, external sources of heat, and electricity. More individuals began to embrace the outdoors and cold environments, as unique terrain such as snow and ice provided heterogeneity to seasons, and people elected to spend more time recreationally in cold environments [10, 11]. This

M. R. Hutchinson, S. Rocha Piedade (eds.), *Sports Medicine: It's Environmental*,
https://doi.org/10.1007/978-3-031-84493-5_3

transition, in addition to growing popularity of the Winter Olympic Games, led to increased participation in outdoor activity and winter sports through the years [10].

Some of these winter sports include cross-country skiing, downhill skiing, snowboarding, long- distance cold-weather running, sled dog racing, and winter mountaineering [12, 13]. With this increased participation likely comes increased exposure to cold-induced injuries [14]. Thus, the sports medicine physician should be well equipped to handle these unique cold-exposure injuries, especially for a growing number of athletes.

3.2 Case Anecdote/Challenge

A 28-year-old male runner training for a winter marathon begins a 10-kilometer run in Greenland three hours before sundown. He is wearing three layers of clothes, as temperature is 5 degrees Celsius and wind strength is 18 miles per hour. While running up a steep incline, he loses his footing on a patch of ice and falls 25 feet down an embankment. Immediately, he feels intense pain in his right lower extremity and is unable to bear weight. He tries to call for help, but his phone has no signal.

He attempts to scoot out of the embankment, but his arms begin to fatigue, and he has ongoing severe pain in his right leg. After 1.5 h, he has increased shivering and some general confusion. His temperature is 34 degrees Celsius. He has an additional long-sleeve shirt in his backpack that he dons and tries to scoot onto the main trail, but there are no bystanders around.

The sun begins to set, and he notices he is no longer shivering. He instead feels more lightheaded with generalized weakness. His temperature is now 30 degrees Celsius. He curls up to keep his core warm, and a bystander finds him in this position.

Initial vitals at this time were a temperature of 30 degrees Celsius, heart rate of 45 beats per minute, respiratory rate of 11 breaths per minute, and blood pressure of 93/60 mmHg. The bystander calls 911 and shares additional layers of clothing to help keep him warm. About 20 min later, first responders arrive to the scene via helicopter. Vitals are rechecked: temperature of 27 degrees Celsius, heart rate of 40 beats per minute, respiratory rate of 6 breaths per minute, and blood pressure of 84/55. EKG is notable for sinus bradycardia (40 beats/minute), first-degree atrioventricular block (PR interval 0.23 s), and J waves at the junction of the QRS and ST segments (Fig. 3.1).

First responders obtain intravenous access and initiate internal rewarming with delivery of heated saline, in addition to external warming via additional blankets and temporary hot packs. Vitals are rechecked every five minutes, with slow improvement in bradycardia and respiratory rate. His cervical spine is stabilized, and his right lower extremity is immobilized prior to transfer on a spine board. He is airlifted to the closest level 1 trauma center. He is diagnosed with an acute right tibial plateau fracture, several right rib fractures, a closed pelvic ring fracture, a small left subdural hematoma, and stage 2 frostbite of right toe digits 1–3. He ends up on extracorporeal life support (ECLS) for several days and undergoes open reduction and internal fixation for his right tibial plateau fracture. He is eventually discharged to acute inpatient rehabilitation prior to discharge home.

This case demonstrates the wide spectrum of management considerations for athletes with prolonged cold exposure and highlights the role of first responders in quickly triaging these patients.

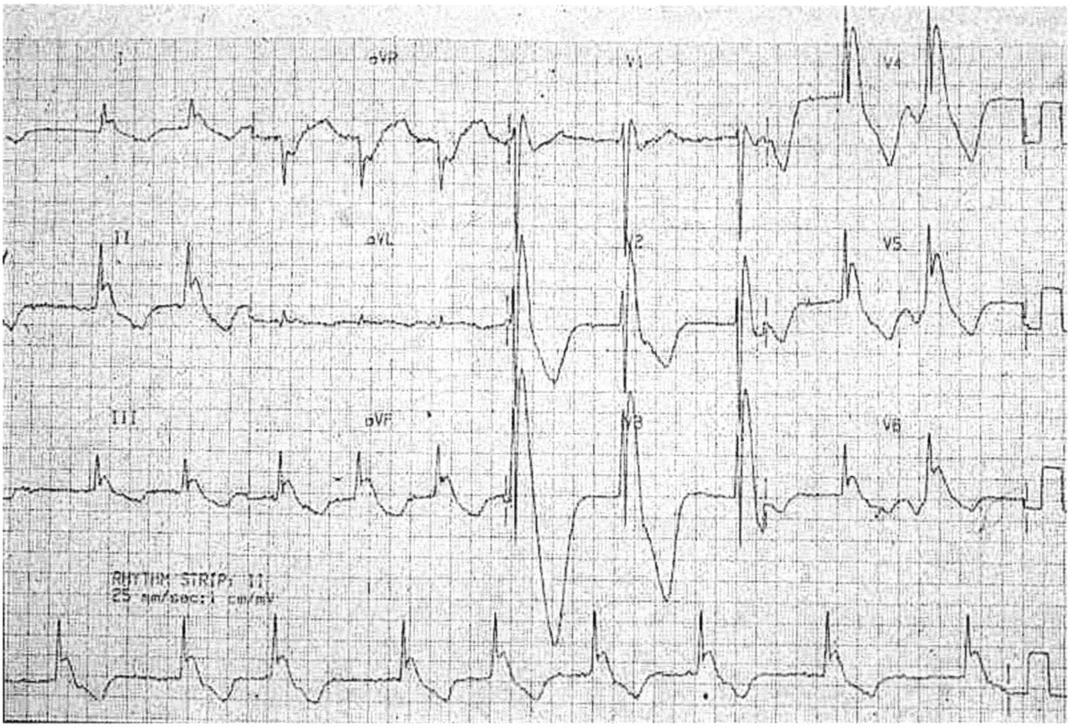

Fig. 3.1 EKG in severely hypothermic patient, with J waves at the junction of QRS and ST segments. (Reproduced with permission from https://litfl.com under the terms of the Creative Commons Attribution-NonCommercial-ShareAlike 4.0 International License)

3.3 Epidemiology: Both in General and Within Sports Medicine

There are many challenges to accurate reporting of cold-induced injuries due to lack of timely presentation to clinical care and geographical barriers such as high-altitude, treacherous environments that preclude identification or rescue of individuals in crisis [15]. Worldwide data on cold-induced injuries is likely underreported due to these inherent challenges [3].

Hypothermia, defined as a core body temperature of <35 degrees Celsius, is one of the most common cold-induced injuries and affects 4 out of 1,000,000 cases in the United States and 1–5 per 100,000 in Europe [16–19]. Mortality rates from hypothermia ranged from 0.2 to 1.81 per 100,000 cases, with rates in the United States lower in general (~2%) [16, 20, 21]. In Canada, a 35-year review of search and rescue data revealed hypothermia as the most common atraumatic medical diagnosis requiring clinical attention, accounting for 49.4% of injuries [22].

Unlike hypothermia, frostbite requires prolonged exposure to subfreezing temperatures. Frostbite, or injury to peripheral digits or extremities from ice crystal formation in subcutaneous tissue, can affect about 1.5 cases per 100,000 people [18]. There is a greater predisposition for those spending prolonged time in colder environments, including the homeless population, those who work in refrigerated environments or outdoors without easy access to warmth, rural populations, northern environments, or extreme altitude winter sports athletes, with risk factors dependent upon environmental, mechanical, physiological, or psychological issues [23] (Table 3.1).

Table 3.1 Risk factors that predispose to cold injury

Climate	**Individual characteristics and physiology**
Wind	African-American or Afro-Caribbean Ethnic Background
Wetness, immersion in cold water	Male
Contact with cold materials	Children
High altitude	Elderly
Hypoxia	Previous cold injury
Long duration or high amount of cold exposure	Poor CIVD response
	Homeless
Behavioral	**Individual/clinical**
Smoking	Coronary artery disease/ischemic heart disease, cardiac insufficiency, stroke
Alcohol use	Peripheral vascular disease
Drug use	Peripheral neuropathy
Inappropriate or constrictive clothing	Cold sensitivity
Prolonged stationary situation, immobility	Raynaud's phenomenon, white fingers
Fatigue, dehydration, malnutrition	Hand-arm-vibration syndrome, vibration
Use of emollients	Diabetes
Military rank/task	Hypothyroidism, hypopituitarism
Poor calorie intake	Depression, schizophrenia, dementia
	Neurovascular diseases
	Sweating or hyperhidrosis
	Previous cold injury
	Medication (vasoconstrictors)

Reproduced with permission from Ref. [27]

Nonfreezing cold injuries, which were initially described during World War I at the peak of trench warfare tactics, occurred because soldiers had prolonged exposure to cold, wet environments, especially with damp socks that led to "trench foot" [24]. Outside of trench warfare, "trench foot" can still occur in some military populations or in homeless populations due to prolonged exposure to wet and cold environments. Another nonfreezing cold injury called chilblains, which involves digit exposure to cold and wet conditions for several hours, can lead to small, erythematous papule formation with pruritus and burning sensations. This can affect athletes such as divers and ice skaters [25, 26].

Risk factors that predispose civilians, military service members, and athletes to all types of cold-weather injuries include extremes of age [27], previous cold injury [28], prolonged outdoor exposure or cold water submersion, inadequate clothing [29], fatigue/exhaustion [27], dehydration, and poor appetite or caloric intake [27]. In general, these can be grouped into categories involving decreased heat production, increased heat loss, or impaired thermoregulation [30]. Older age over 60 years is usually associated with decreased adiposity and higher risks of cardiovascular disease, which makes the natural process of peripheral vasoconstriction to maintain warmth somewhat dangerous. Younger age can impact cold injury prevalence rates due to lack of education on maintaining proper body temperature and decreased subcutaneous tissue for insulation [27]. In military literature, soldiers were 1.68 times more likely to sustain a repeat cold-related injury if they had one previously [31]. Poor caloric intake or nutritional deficiencies prevent the body's internal glycolytic fuel stores from promoting thermogenesis via shivering for prolonged periods of time, accelerating risk of organ failure. After exercising to exhaustion, with subsequent cold exposure, core body temperature is reduced significantly; thus, exercise to exhaustion, which is often the case in elite athletes, may predispose to hypothermia [27]. Reports of gender predisposing to cold-related injuries are less concrete since military data skews toward a higher male population at baseline. However, the US Armed Forces notes being female as a risk factor, likely due to decreased total body mass and decreased muscle mass [27]. In general, higher percent body fat and higher muscle mass promote better insulation of core body temperature and protect against cold-induced injuries [27].

Medical risk factors that compound cold-induced injury include a history of smoking; chronic vascular disease, such as peripheral artery disease, peripheral vascular disease, or diabetes,; or medical history requiring medications that decrease sympathetic response (such as beta-blockers) [32, 33]. These are risk factors due to the inherent homeostatic mechanisms to maintain core body temperature by inducing peripheral vasoconstriction, which exacerbates risk for frostbite or poor circulation for patients with significant vascular disease or on medications that promote peripheral vasoconstriction. Conditions leading to impaired sensation, such as diabetes, also increase risk [34].

In more recent years, as recreational activities in cold-weather environments have been more popular, mountain sports have had a higher rate of injuries due to prolonged time in cold, high-altitude environments [35]. Hiking, trekking, and downhill skiing were rated some of the most popular mountain sports, with activities such as ski mountaineering and snowshoeing also becoming more popular recently [36]. One study reported that rates of frostbite and hypothermia could affect up to 20% of athletes, whereas frostbite in mountaineers affected 366 per 1000 individuals per year in a cross-sectional study over two years [29, 37]. It was thought that external sources were most at play for increasing cold-related injuries in mountaineers, including inappropriate clothing or equipment [29]. In dog sledding, data from the 2010 Iditarod dog sled race illustrated that frostbite was the most prevalent injury, affecting almost one-third of participants [12]. Overall, in previous years, the highest risk event at the Olympics for hypothermia was open-water swimming given higher duration of time spent in the water, which accelerates loss of heat from the body; otherwise, most Olympic sports are shorter in duration and generate high enough metabolic demand to mitigate risk for cold injury [23, 38].

3.4 Unique Injuries/Illnesses/Patterns

Unique injuries from prolonged exposure to cold can induce a variety of illness patterns. One of the most common cold-induced injuries, hypothermia, occurs when the core body temperature dips below 35 degrees Celsius [27]. Hypothermia can be divided into several categories, mild versus moderate versus severe, based on body temperature. The body has a unique physiologic response to severity of cold based on core body temperature, from shivering to maintain body warmth to shock/asystole in severe cases (Table 3.2).

Table 3.2 Gradation of severity of hypothermia with associated symptoms based on temperature

Mild (32–35 °C; 89.6–95 °F)	Moderate (28–31 °C; 82.4–88 °F)	Severe (<28 °C; <82.4 °F)
Cold extremities	Apathy, poor judgment	Inappropriate behavior
Shivering	Reduced shivering	Total loss of shivering
Tachycardia	Weakness and drowsiness	Cardiac arrhythmias
Tachypnea	Slurred speech and amnesia	Pulmonary edema
Urinary urgency	Dehydration	Hypotension and bradycardia
Mild incoordination	Decreased coordination or clumsiness	Reduced level of consciousness
	Fatigue	Muscle rigidity

Reproduced with permission from Ref. [27]

Wind (mph)	Temperature (°F)																	
Calm	40	35	30	25	20	15	10	5	0	-5	-10	-15	-20	-25	-30	-35	-40	-45
5	36	31	25	19	13	7	1	-5	-11	-16	-22	-28	-34	-40	-46	-52	-57	-63
10	34	27	21	15	9	3	-4	-10	-16	-22	-28	-35	-41	-47	-53	-59	-66	-72
15	32	25	19	13	6	0	-7	-13	-19	-26	-32	-39	-45	-51	-58	-64	-71	-77
20	30	24	17	11	4	-2	-9	-15	-22	-29	-35	-42	-48	-55	-61	-68	-74	-81
25	29	23	16	9	3	-4	-11	-17	-24	-31	-37	-44	-51	-58	-64	-71	-78	-84
30	28	22	15	8	1	-5	-12	-19	-26	-33	-39	-46	-53	-60	-67	-73	-80	-87
35	28	21	14	7	0	-7	-14	-21	-27	-34	-41	-48	-55	-62	-69	-76	-82	-89
40	27	20	13	6	-1	-8	-15	-22	-29	-36	-43	-50	-57	-64	-71	-78	-84	-91
45	26	19	12	5	-2	-9	-16	-23	-30	-37	-44	-51	-58	-65	-72	-79	-86	-93
50	26	19	12	4	-3	-10	-17	-24	-31	-38	-45	-52	-60	-67	-74	-81	-88	-95
55	25	18	11	4	-3	-11	-18	-25	-32	-39	-46	-54	-61	-68	-75	-82	-89	-97
60	25	17	10	3	-4	-11	-19	-26	-33	-40	-48	-55	-62	-69	-76	-84	-91	-98

Frostbite Times: 30 minutes; 10 minutes; 5 minutes

Fig. 3.2 Effect of wind (in miles per hour) on temperature and time to frostbite (in minutes). (Reproduced with permission from Ref. [27])

Prolonged exposure to subfreezing temperatures can cause frostbite or tissue damage primarily to the digits or nose because of ice crystal formation in the skin or subcutaneous tissue [27]. Frostbite occurs on a spectrum, from superficial (first or second degree, causing little to no tissue loss) to deep (third or fourth degree, marked by tissue loss) [27, 30]. Exposure to these subfreezing temperatures can be compounded by the effect of wind in causing lower core body temperature; a wind-chill index is used as to determine a person's exposure/response to cold and how quickly frostbite can occur [30] (Fig. 3.2).

In addition to risks associated by sheer nature of exposure to cold environments, multiple medical comorbidities can predispose to cold-related illness patterns affecting multiple organ systems, outside of typical freezing/nonfreezing cold injuries, especially during participation in sport.

From a pulmonary standpoint, exercise-induced bronchospasm (EIB) is exacerbated by prolonged exposure to cold. After an athlete inhales, the nose is used as a humidifying and warming agent for air passage to the lungs; however, in freezing environments, the nose is unable to adequately and consistently warm air. Upon contact of dry, cold air within the bronchioles, a reflexive bronchoconstriction occurs, given increased osmolality along the airway surface lining, to increase minute ventilation, which decreases overall exercise tolerance [39]. This leads athletes to report chest tightness, dyspnea on exertion, or persistent cough/wheezing. Furthermore, this increase in minute ventilation causes athletes to revert to mouth-breathing, especially since rhinorrhea is more prominent in the cold, which bypasses the nose as an initial warming agent in the first place [40]. With increased bronchoconstriction yields increased inflammatory mediators within the bronchioles, triggering mast cell activation and leading to reactive airway edema which can exacerbate dyspnea. This well-documented effect has affected those participating in sports with high velocity in cold environments, including speed skaters [41], ice hockey [42], Nordic sports [43], and long-distance runners [44]. There was a total incidence of 23% of athletes with EIB in the 1998 US Winter Olympics, with highest percentages reported in the cross-country skiers [41]. Those with underlying asthma are even at higher risk of EIB with cold-exercise participation in sport.

From a cardiovascular standpoint, decreased skin temperature typically leads to peripheral vasoconstriction to improve cardiac output and central venous return. Due to this increased myocardial oxygen consumption to maintain homeostasis, anginal thresholds are lowered, thus increasing the rate of sudden ischemic cardiac death. A study in a known cold-weather region, Finland, showed an increased rate of sudden cardiac death in over 2500 patients associated with sharp, several-day declines in temperature [45]. This can be exacerbated in athletes who are also increasing cardiac workload to exercise. Furthermore, prolonged exposure to cold with body temperatures below 35 degrees Celsius can lead to cardiac arrhythmias including atrial fibrillation, ventricular fibrillation, or widening of the QRS complex and QT interval [46]. These long-term effects of cardiovascular decline are more prominent in cold water immersion sports such as open water or winter swimming [47, 48].

From a musculoskeletal standpoint, muscle strength and flexibility/range of motion can be compromised in cold weather, leading to overall decreases in power, compliance, and endurance [49]. There are increased reports of musculoskeletal neck and back pain for those who work in colder environments [50]. There is also higher reported knee injury risk in alpine skiers who spend more time in cold air, which suggests that cold can exacerbate musculoskeletal pain [51].

From a skin standpoint, more minor, non-life-threatening cold-induced injuries include cold urticaria, which affects about 0.05% of the population. Mast cells within the skin activate and release histamine after exposure to cold, which can cause local, often pruritic rashes or wheals. Remaining in the cold increases risk of progression to anaphylactic shock; thus, athletes experiencing these symptoms should carry an EpiPen at all times [52]. Another condition called chilblains, or digit exposure to cold and wet conditions for several hours, can occur in skaters or long-distance swimmers [25].

Raynaud's phenomenon, or peripheral vasoconstriction in response to cold, affects an estimated 3–5% of the population and can cause discoloration or digit ischemia. This can occur in isolation or due to an underlying rheumatologic or vascular disorder. Athletes with Raynaud's phenomenon should have a full rheumatologic and vascular workup to ensure no additional primary disorder, and these athletes should not participate in sports with prolonged cold exposure due to risk of auto-amputation [53]. A similar condition called cryoglobulinemia vasculitis can affect individuals when body temperature reaches below 37 degrees Celsius, and small purpura appear with associated weakness, edema, or skin ulcerations. This is usually discovered as a sequelae of a separate primary diagnosis, so treatment is typically targeted toward the primary diagnosis [54].

Overall, participation in cold-weather related sports can be safe with proper precautions but is not without risk. The sports medicine physician should be cognizant of the multiorgan effect that cold can have on athletes before, during, and after activity.

3.5 Management

Management of cold-related injuries can be difficult to navigate given unpredictable climate, altitude differences, treacherous terrain, and proper identification of someone in need of medical assistance. Prior to managing an athlete in need of medical attention, treating physicians must first ensure the scene is safe for them to avoid further injury to themselves or others [27].

For management of all types of cold injuries, the priority is to remove the athlete from the offending environment as soon as possible and to remove any clothing items that lower core body temperature, such as damp clothing, which promotes evaporative cooling. In land settings, creating a shelter from cold weather or wind can mitigate injury, such as camping tents or tarps [27]. In athletes who are long-distance swimmers, this involves removal from the water and drying of the athlete.

In cases of mild hypothermia (body temperature 32–35 degrees Celsius), passive rewarming via blankets and additional layers, or sharing warm drinks or food, should be trialed to help improve core body temperature. If core body temperature improves to normal levels quickly, escalation of medical care to a hospital is not required. However, in moderate hypothermia (body temperature 28–32 degrees Celsius), a process called active rewarming should begin quickly. External sources of warmth should be directed toward the core instead of the periphery to prevent further vasoconstriction. Frequent skin checks should occur to ensure no development of burns during the warming process [27]. For the athlete, access to peripheral warmth, including blankets, heat packs, warmers, extra clothes, and mittens, should be at all major cold-weather events such as the Winter Olympics.

For those with severe hypothermia (body temperatures <28 degrees Celsius), both internal and external warming are required; however, internal warming can be difficult based on terrain/access. Heated isotonic saline (to 40–42 degrees Celsius) can be administered intravenously for internal warming, at a rate that increases body temperature from 0.75 to 1 degrees Celsius per hour [27]. Heat packs, blankets, and warmers can be used as external sources of heat. Athletes should avoid bearing weight if possible, to avoid exacerbating underlying traumas or to prevent avulsion fractures, and ideally would be transported immobilized in case of undiagnosed traumatic injuries [27].

Furthermore, there are basic and advanced life support considerations that are unique to treating a hypothermic athlete. When performing advanced cardiac life support (ACLS) resuscitation on a hypothermic patient, medications such as epinephrine should not be administered until the patient reaches a core body temperature of at least 30 degrees Celsius due to increased cardiac strain with these medications. The number of defibrillations recommended in the hypothermic patient varies based on governing body; the European guidelines recommend three defibrillation shocks at maximum Joules if body temperature is <30 degrees Celsius, and extra caution should be taken when assessing vitals due to higher risk of bradycardia and decreased respirations in the cold [55].

For patients suffering from frostbite, removal from the offending environment should again be the priority, and it should be determined if there is a possibility of preventing a refreezing injury after an initial thaw is attempted [56]. Repeated refreezing and thawing in this setting can lead to underlying damage to tissues. If there is high likelihood that the individual can be removed from the cold environment, an attempt to externally warm should be attempted; possibilities include immersion into a heated, circulating water bath at 37–39 degrees Celsius until the skin is either red/purple and soft [56], application of multiple layers of insulated clothing to a targeted site, or use of a Bair Hugger. Direct contact with heaters, fire, or coal should be avoided given high risk of burns and decreased sensory impulses in cold. As an intermediate tool, first responders can use their own body heat to share warmth, with body parts like the axilla typically being the warmest [57]. A theoretical administration of a nonsteroidal anti-inflammatory medication (NSAIDs) has been discussed, since NSAIDs promote prostaglandin release, inhibit proinflammatory cytokines, and decrease vasoconstriction. In some hospitals, use of iloprost (a synthetic prostacyclin analog) within 72 hours of initial injury can reduce risk of amputation in grades 3 and 4 frostbite [58]. A summary of management options starting from initial evaluation of an athlete with a cold-induced injury can be seen in Fig. 3.3.

For patients with cold water immersion injuries, such as long-distance swimmers, water should be dried from the skin as soon as possible upon evacuation to mitigate effects of evaporative cooling. Water has a significantly higher thermal capacity than air, thereby promoting convective heat loss from the surface of the skin; the higher the total body surface area a person is submerged, the lower their core body temperature will be [27].

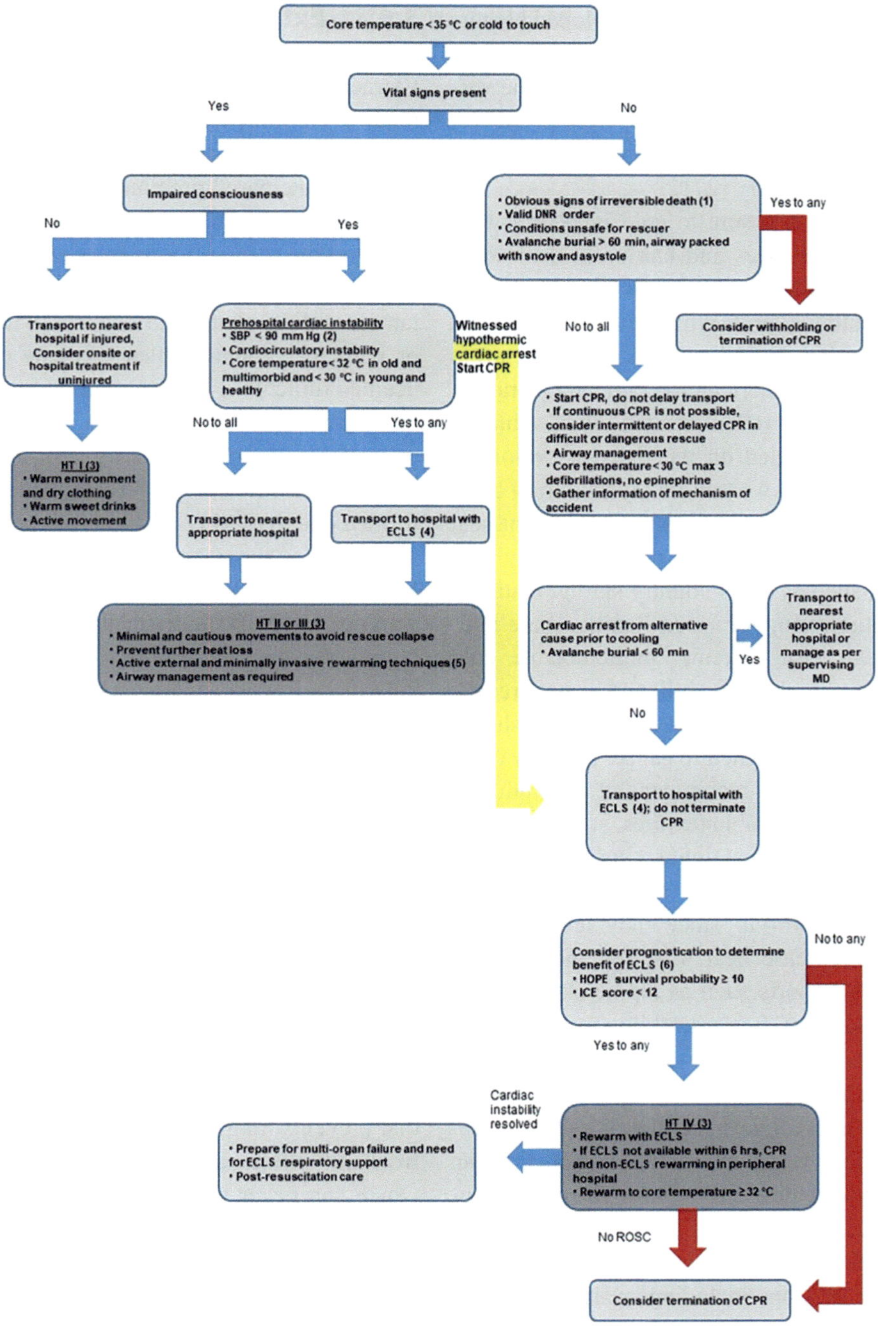

Fig. 3.3 Proposed management of an acutely hypothermic athlete in the field, from initial evaluation to hospital course. (Reproduced with permission from Ref. [27])

If an individual has already developed blisters or vesicles from prolonged cold exposure, dry gauze should be applied loosely to cover the vesicles, but dressing should not be tight given risk for increasing edema or minimizing blood circulation to the periphery [27].

Sports with exposure to cold involve a wide spectrum of activities and can be broken down into four major categories, from low risk to high risk. For outdoor sports with occasional exposure to cold weather, including track and field, cross country, soccer, American football, and marathon canoeing, it is recommended that athletes/coaches be educated on the risks of prolonged cold exposure and wear more layers during competition that wick away moisture, to prevent heat loss. For endurance sports that are primarily winter based, such as cross-country skiing, biathlon, ski mountaineering, or speed skating, these individuals can face extreme, unpredictable cold temperatures, high wind chill, and heavy precipitation. The governing bodies of each sport should have guidelines on when to cancel or delay competition. It is also recommended that athletes layer up with breathable textures, such as merino wool, instead of typical tight racing suits that provide minimal warmth, and to change base layers after initial warm-up since they are likely to sweat and increase heat loss. For more speed-based winter sports such as alpine skiing, luge/bobsled, skeleton, freestyle skiing, and downhill mountain biking, the additional elements of wind with high speeds play an even larger factor, which accelerate convective heat loss, so similar precautions are in line about multiple layers that can wick away moisture and to improve general education of athletes/coaches and trainers involved in care [59]. For cold water-based sports that require immersion, including triathlons, World Aquatics (FINA) mandates use of a wetsuit for water temperatures below 15 degrees Celsius [59].

3.6 Extraction Challenges

Extraction of an athlete with a cold-related injury requires careful coordination and planning, especially in high-altitude environments with uneven terrain or increased precipitation. The initial priority for the first responder is to always ensure safety of the scene prior to entry.

In the specific scenario of extraction of an avalanche victim, which can occur in winter sports or with recreational athletes, rescue efforts are deemed futile if the victim is buried for more than 35 min with airway obstruction, if electrocardiogram shows no rhythm or sign of life, if lethal injury is identified, or if CPR is unable to be performed due to rigidity of the chest wall [55, 60].

There are three phases for individuals in an avalanche burial: survival, asphyxiation, and waiting. Over 100 individuals die in avalanche burials per year in Europe, with an estimated 25% mortality rate, and most of these occur from individuals participating in recreational events [60, 61]. Longitudinal data from 1981 to 1991 in European skiers suggested that 93% of avalanche victims were still alive in the first 15–20 min; after 30–35 min, >65% die due to freezing temperature and asphyxiation, with the most common cause of death being a combination of hypoxia, hypothermia, and hypercapnia [62]. The cause of death for most (60%) occurs from asphyxiation within the first 35 min of burial, as a combination of lack of a means for airway and dense compression against the thoracic cavity that prevents lung expansion, while 10–20% die from initial trauma [62]. More humid locations, which can occur in coastal areas, lead to dense snow that can accelerate this risk of death in a burial scenario [63].

Factors that help with survival in avalanche burials include decreased depth of avalanche, decreased duration of burial, larger human body/increased muscle mass for core insulation, presence of bystanders, presence of an air pocket, and use of an avalanche airbag [60, 62, 63]. Athletes may be at a slight advantage given typically increased muscle mass from baseline. Avalanche airbags can reduce burial depth for winter athletes such as snowboarders or skiers, and use of a tracking device on each athlete called an avalanche transceiver allows for accurate identification of victim location. This exact geolocation can help identify athletes who become entrapped under the snow. A system called RECCO incorporates geolocation technology to identify victims and is used by over 600 rescue organizations across the world [64]. If this fails, the next step is to initiate a probe-line search, and avalanche rescue dogs can be used to locate victims by scent [64]. There have been increasing rates of helicopter extractions for mountain rescues, as helicopters typically provide a faster and safer response than a coordinated ground rescue, but time for response still takes longer than 20 min [35].

Extrication of an avalanche burial victim begins with identifying victim location and then building a channel toward the head, since highest rates of death occur by asphyxiation [60, 62, 63]. Vital signs, including temperature, are ideally assessed via an esophageal probe or epitympanic thermometer using a thermistor if there is no snow or obstruction of the external auditory canal, and an electrocardiogram should be obtained before transport of the victim, since cardiac arrhythmias can occur with movement [65]. Defibrillator pads should be placed, and CPR should begin with administration of five rescue breaths due to high rates of asphyxiation [55]. Access intravenously or intraosseously should be obtained; due to peripheral vasoconstriction, it is often easier to administer fluids or medications intraosseously [66].

Since avalanche burials can often lead to traumatic injuries, transport of victims requires spinal precautions, administration of analgesics, splinting/immobilization in case of any orthopedic injuries, and minimal movement given risk of precipitating shock. If there is severe head trauma, emergent intubation or use of a bag valve mask for oxygenation is recommended [67]. For those with core body temperatures <30 degrees Celsius, systolic blood pressures <90 mmHg, and cardiac arrhythmias, transportation to an ECLS site should be prioritized [55]. Once patients arrive at the hospital, a Hypothermia Outcome Prediction after ECLS (HOPE) score is calculated to ascertain if a hypothermic patient in cardiac arrest would benefit from ECLS rewarming [68]. The extrication process for an avalanche burial victim is summarized in Fig. 3.4.

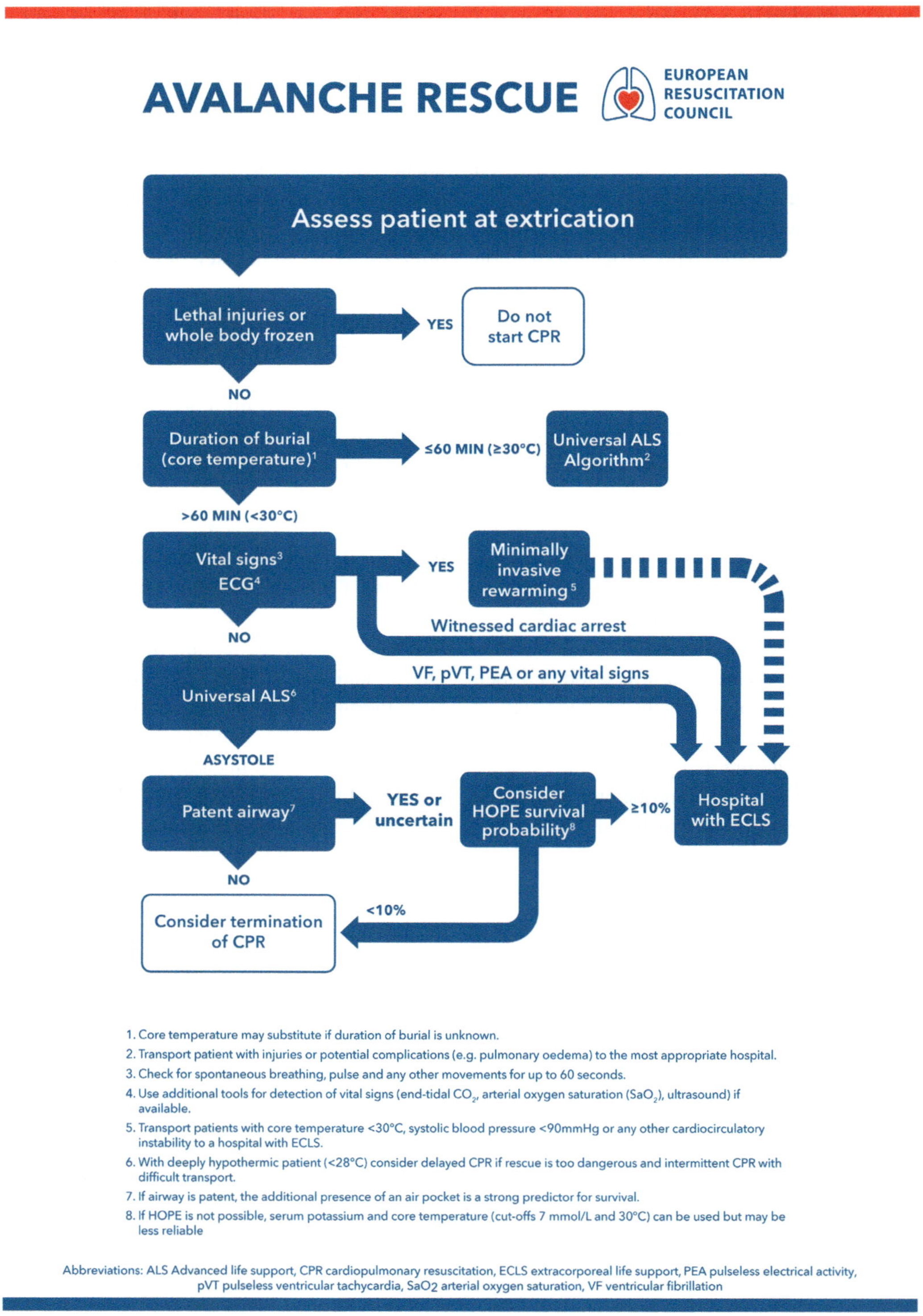

Fig. 3.4 Extrication process for avalanche rescue. (Reproduced with permission from Ref. [55])

3.7 Prevention

The American College of Sports Medicine has a formal recommendation that exercise in cold environments can be safe as long as comprehensive precautions are taken to mitigate injury risk [27]. An approach to prevention of cold-induced injuries requires a thorough assessment of natural conditions prior to competition and ensures athletes have proper equipment to withstand temperature extremes. Prevention strategies should be targeted toward an athlete's medical history and his/her sport.

Prevention strategies for hypothermia include proper education on the detrimental effects of prolonged cold exposure on the body, ensuring appropriate clothing with multiple layers covering as much skin surface area as possible but without peripheral constriction of extremities, or travelling with sources of external heat such as heat packs. The innermost layer should be composed of polypropylene, polyester, or synthetic wool, as these insulating materials can prevent moisture accumulation. The middle layer should be composed of a primary insulating material such as fleece or wool. The outermost layer should also wick away moisture and protect against wind while allowing for proper ventilation [27, 30, 57]. If using extra socks, athletes should size up at least one shoe size to prevent circulation issues [27]. For prevention of trench foot, individuals should change socks 2–3 times per day, with use of vapor barrier boots. Trials of antiperspirant on feet can help decrease sweating in the foot [27]. Alcohol and caffeine intake should be avoided due to causing peripheral vasoconstriction [27, 30]. A summary of preventative measures for cold-induced injuries in athletes can be found in Fig. 3.5.

Outside of modifications to clothing, it is important to note the impact of nutrition/body weight on protection against cold-induced injuries. Ensuring that athletes are maintaining high caloric intake given high energy expenditure to maintain warmth in cold environments will allow for improved mechanisms to maintain warmth, such as shivering [27]. It is generally recommended to consume high carbohydrate foods to help with the quick transfer of energy for usage, especially for athletes [27].

Exercise itself has also been shown to increase metabolic rate and core body temperature. If concern for caloric depletion or exhaustion is less of an issue, exercise can help prevent frostbite from occurring in dry conditions with lack of wind [27]. To promote these preventative behaviors, it is recommended that athletes not travel solo when participating in sports that require prolonged cold exposure and perform routine skin checks to monitor for early manifestations of frostbite [27].

Prevention of more unique effects of cold-induced injuries on the body, including EIB, can involve medication or non-medication management. Short-acting beta agonists should be used 15 min before the start of exercise to help mitigate EIB. Furthermore, ensuring proper warm-up activities for at least 15 min at a moderate intensity can be helpful, in addition to attempts to humidify air via a face mask [69].

In general pre-participation physicals for athletes at risk for cold-induced injuries, additional questions regarding an athlete's weight, diet, history of prior cold-induced injuries, general education on cold-induced injuries, cardiovascular history, history of thyroid problems, and history of rheumatologic problems should be assessed given known risk factors predisposing to morbidity [30].

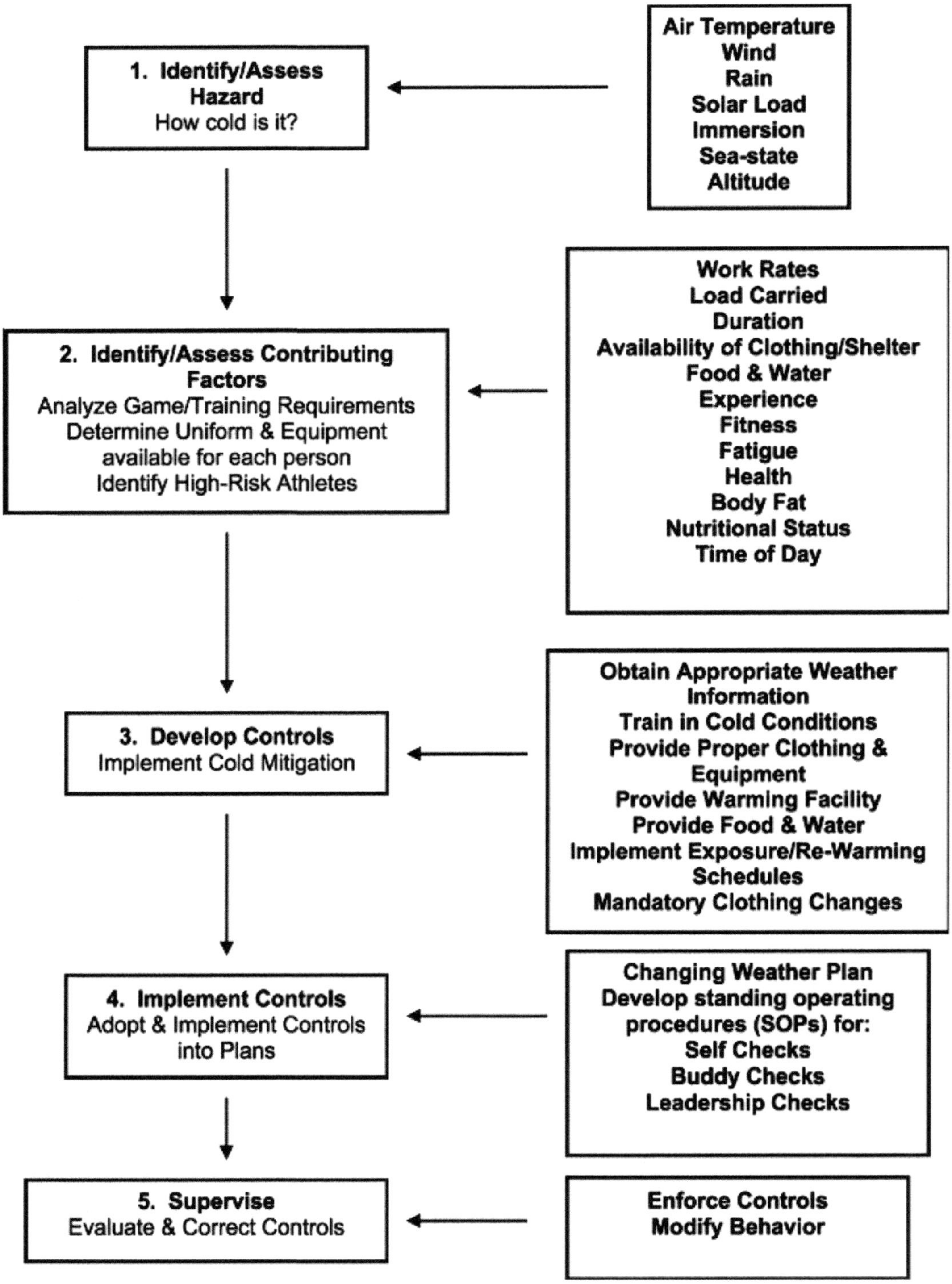

Fig. 3.5 Proposed workflow to prevent cold injuries in athletes. (Reproduced with permission from Ref. [70])

3.8 Take-Home Pearls

- Athletes, especially those participating in winter sports or outdoor immersion sports, can be uniquely predisposed to cold-induced injuries based on sustained exposure to cold and/or windy environments without adequate protection, even in nonfreezing temperatures.
- The most common cold injuries include hypothermia, which can be divided into mild (body temperature 32–35 degrees Celsius) versus moderate (body temperature 28–32 degrees Celsius) versus severe (<28 degrees Celsius), with worsening clinical symptoms and greater risk for morbidity/mortality at colder temperatures.
- Athletes are at high risk for a range of frostbite injuries based on time of exposure in subfreezing temperatures; early identification and prevention is key in reducing negative long-term outcomes such as amputation.
- Athletes with underlying asthma and cardiovascular or rheumatologic disease should undergo additional screening and understand precautions of exercise in cold environments.
- Management of hypothermia requires rapid identification of an individual with reduced core body temperature, application of warmth safely, and extrication as soon as possible. There are specific considerations for ACLS protocols on managing the hypothermic patient.
- Prevention of hypothermia requires robust education of athletes, teams, and coaches on symptoms concerning for cold-induced injury and active involvement by governing bodies to postpone/cancel competitions based on standardized temperature/wind chill data to keep athletes safe.

References

1. Enander A, Ljungberg AS, Holmer I. Effects of work in cold stores on man. Scand J Work Environ Health. 1979;5(3):195–204.
2. Ghani N, Tariq F, Javed H, Nisar N, Tahir A. Low-temperature health hazards among workers of cold storage facilities in Lahore. Pak Med Pr. 2020;71(1):1–7.
3. Paal P, Pasquier M, Darocha T, Lechner R, Kosinski S, Wallner B, et al. Accidental Hypothermia: 2021 update. Int J Environ Res Public Health. 2022;19(1)
4. Castellani JW, Young AJ. Human physiological responses to cold exposure: acute responses and acclimatization to prolonged exposure. Auton Neurosci. 2016;196:63–74.
5. Burtscher M, Gatterer H, Burtscher J, Mairbaurl H. Extreme terrestrial environments: life in thermal stress and hypoxia. A narrative review. Front Physiol. 2018;9:572.
6. Bristow GK, Giesbrecht GG. Contribution of exercise and shivering to recovery from induced hypothermia (31.2 degrees C) in one subject. Aviat Space Environ Med. 1988;59(6):549–52.
7. Eyolfson DA, Tikuisis P, Xu X, Weseen G, Giesbrecht GG. Measurement and prediction of peak shivering intensity in humans. Eur J Appl Physiol. 2001;84(1–2):100–6.
8. Friedman LS, Abasilim C, Fitts R, Wueste M. Clinical outcomes of temperature related injuries treated in the hospital setting, 2011–2018. Environ Res. 2020;189:109882.
9. Makinen TM. Different types of cold adaptation in humans. Front Biosci (Schol Ed). 2010;2(3):1047–67.
10. Zhang W, Li H, Wang D, Xu G, Xu C, Li J, et al. The global research status and trends in ice and snow sports injuries from 1995 to 2022: a bibliometric and visualized analysis. Int J Environ Res Public Health. 2023;20(4)
11. Vogel K, Hulsopple C. Cold weather injuries: initial evaluation and management. Curr Sports Med Rep. 2022;21(4):117–22.
12. Gallea JW, Higgins GL 3rd, Germann CA, Strout TD. Injury and illness sustained by human competitors in the 2010 iditarod sled dog race. Am J Emerg Med. 2014;32(7):780–4.
13. Zang W, Fang M, Zhang X, Xiao N, Wang S, Mu L. Exploring the epidemiology of injuries in athletes of the Olympic winter games: a systematic review and meta-analysis. J Sports Sci Med. 2023;22(4):748–59.
14. Wu P, Zhu X, Yang S, Huang J. The influence of the Beijing Winter Olympic games on the demand for winter sports: an empirical analysis based on the Baidu Index. Heliyon. 2023;9(10):e20426.
15. Wallner B, Moroder L, Brandt A, Mair P, Erhart S, Bachler M, et al. Extrication times during avalanche companion rescue: a randomized single-blinded Manikin study. High Alt Med Biol. 2019;20(3):245–50.
16. Baumgartner EA, Belson M, Rubin C, Patel M. Hypothermia and other cold-related morbidity emergency department visits: United States, 1995–2004. Wilderness Environ Med. 2008;19(4):233–7.
17. Kosinski S, Darocha T, Galazkowski R, Drwila R. Accidental hypothermia in Poland – estimation of

prevalence, diagnostic methods and treatment. Scand J Trauma Resusc Emerg Med. 2015;23:13.
18. Brandstrom H, Johansson G, Giesbrecht GG, Angquist KA, Haney MF. Accidental cold-related injury leading to hospitalization in northern Sweden: an eight-year retrospective analysis. Scand J Trauma Resusc Emerg Med. 2014;22:6.
19. Bierens JJ, Uitslager R, Swenne-van Ingen MM, van Stiphout WA, Knape JT. Accidental hypothermia: incidence, risk factors and clinical course of patients admitted to hospital. Eur J Emerg Med. 1995;2(1):38–46.
20. Herity B, Daly L, Bourke GJ, Horgan JM. Hypothermia and mortality and morbidity. An epidemiological analysis. J Epidemiol Community Health. 1991;45(1):19–23.
21. Taylor NA, Griffiths RF, Cotter JD. Epidemiology of hypothermia: fatalities and hospitalisations in New Zealand. Aust NZ J Med. 1994;24(6):705–10.
22. Curran-Sills GM, Karahalios A. Epidemiological trends in search and rescue incidents documented by the Alpine Club of Canada From 1970 to 2005. Wilderness Environ Med. 2015;26(4):536–43.
23. Castellani JW, Young AJ. Health and performance challenges during sports training and competition in cold weather. Br J Sports Med. 2012;46(11):788–91.
24. Mistry K, Ondhia C, Levell NJ. A review of trench foot: a disease of the past in the present. Clin Exp Dermatol. 2020;45(1):10–4.
25. Tlougan BE, Mancini AJ, Mandell JA, Cohen DE, Sanchez MR. Skin conditions in figure skaters, ice-hockey players and speed skaters: Part I – mechanical dermatoses. Sports Med. 2011;41(9):709–19.
26. Laden GD, Purdy G, O'Rielly G. Cold injury to a diver's hand after a 90-min dive in 6 degrees C water. Aviat Space Environ Med. 2007;78(5):523–5.
27. Castellani JW, Eglin CM, Ikaheimo TM, Montgomery H, Paal P, Tipton MJ. ACSM expert consensus statement: injury prevention and exercise performance during cold-weather exercise. Curr Sports Med Rep. 2021;20(11):594–607.
28. Cattermole TJ. The epidemiology of cold injury in Antarctica. Aviat Space Environ Med. 1999;70(2):135–40.
29. Harirchi I, Arvin A, Vash JH, Zafarmand V. Frostbite: incidence and predisposing factors in mountaineers. Br J Sports Med. 2005;39(12):898–901; discussion
30. Fudge J. Preventing and Managing Hypothermia and Frostbite Injury. Sports Health. 2016;8(2):133–9.
31. Sumner DS, Criblez TL, Doolittle WH. Host factors in human frostbite. Mil Med. 1974;141(6):454–61.
32. Biem J, Koehncke N, Classen D, Dosman J. Out of the cold: management of hypothermia and frostbite. CMAJ. 2003;168(3):305–11.
33. Imray CH, Oakley EH. Cold still kills: cold-related illnesses in military practice freezing and non-freezing cold injury. J R Army Med Corps. 2005;151(4):218–22.
34. Rathjen NA, Shahbodaghi SD, Brown JA. Hypothermia and Cold Weather Injuries. Am Fam Physician. 2019;100(11):680–6.
35. Kaufmann M, Moser B, Lederer W. Changes in injury patterns and severity in a helicopter air-rescue system over a 6-year period. Wilderness Environ Med. 2006;17(1):8–14.
36. Procter E, Strapazzon G, Dal Cappello T, Castlunger L, Staffler HP, Brugger H. Adherence of backcountry winter recreationists to avalanche prevention and safety practices in northern Italy. Scand J Med Sci Sports. 2014;24(5):823–9.
37. Hixson EG. Injury patterns in cross-country skiing. Phys Sportsmed. 1981;9(12):44–53.
38. Castro RR, Mendes FS, Nobrega AC. Risk of hypothermia in a new Olympic event: the 10-km marathon swim. Clinics (Sao Paulo). 2009;64(4):351–6.
39. Kennedy MD, Steele AR, Parent EC, Steinback CD. Cold air exercise screening for exercise induced bronchoconstriction in cold weather athletes. Respir Physiol Neurobiol. 2019;269:103262.
40. Silvers WS. The skier's nose: a model of cold-induced rhinorrhea. Ann Allergy. 1991;67(1):32–6.
41. Wilber RL, Rundell KW, Szmedra L, Jenkinson DM, Im J, Drake SD. Incidence of exercise-induced bronchospasm in Olympic winter sport athletes. Med Sci Sports Exerc. 2000;32(4):732–7.
42. Rundell KW, Spiering BA, Evans TM, Baumann JM. Baseline lung function, exercise-induced bronchoconstriction, and asthma-like symptoms in elite women ice hockey players. Med Sci Sports Exerc. 2004;36(3):405–10.
43. Rundell KW, Spiering BA, Judelson DA, Wilson MH. Bronchoconstriction during cross-country skiing: is there really a refractory period? Med Sci Sports Exerc. 2003;35(1):18–26.
44. Ucok K, Dane S, Gokbel H, Akar S. Prevalence of exercise-induced bronchospasm in long distance runners trained in cold weather. Lung. 2004;182(5):265–70.
45. Ryti NR, Makikyro EM, Antikainen H, Junttila MJ, Hookana E, Ikaheimo TM, et al. Cold spells and ischaemic sudden cardiac death: effect modification by prior diagnosis of ischaemic heart disease and cardioprotective medication. Sci Rep. 2017;7:41060.
46. Cappaert TA, Stone JA, Castellani JW, Krause BA, Smith D, Stephens BA, et al. National Athletic Trainers' Association position statement: environmental cold injuries. J Athl Train. 2008;43(6):640–58.
47. Tipton M, Bradford C. Moving in extreme environments: open water swimming in cold and warm water. Extrem Physiol Med. 2014;3:12.
48. Manolis AS, Manolis SA, Manolis AA, Manolis TA, Apostolaki N, Melita H. Winter swimming: body hardening and cardiorespiratory protection via sustainable acclimation. Curr Sports Med Rep. 2019;18(11):401–15.

49. Wakabayashi H, Matsumoto K, Kobori Y, Ebara T, Matsushita M, Kameya T, et al. Multiorgan contribution to non-shivering and shivering thermogenesis and vascular responses during gradual cold exposure in humans. Eur J Appl Physiol. 2020;120(12):2737–47.
50. Hildebrandt VH, Bongers PM, van Dijk FJ, Kemper HC, Dul J. The influence of climatic factors on non-specific back and neck-shoulder disease. Ergonomics. 2002;45(1):32–48.
51. Ruedl G, Fink C, Schranz A, Sommersacher R, Nachbauer W, Burtscher M. Impact of environmental factors on knee injuries in male and female recreational skiers. Scand J Med Sci Sports. 2012;22(2):185–9.
52. Krause K, Zuberbier T, Maurer M. Modern approaches to the diagnosis and treatment of cold contact urticaria. Curr Allergy Asthma Rep. 2010;10(4):243–9.
53. Wigley FM, Flavahan NA. Raynaud's Phenomenon. N Engl J Med. 2016;375(6):556–65.
54. Muchtar E, Magen H, Gertz MA. How I treat cryoglobulinemia. Blood. 2017;129(3):289–98.
55. Lott C, Truhlar A, Alfonzo A, Barelli A, Gonzalez-Salvado V, Hinkelbein J, et al. European Resuscitation Council Guidelines 2021: cardiac arrest in special circumstances. Resuscitation. 2021;161:152–219.
56. McIntosh SE, Opacic M, Freer L, Grissom CK, Auerbach PS, Rodway GW, et al. Wilderness Medical Society practice guidelines for the prevention and treatment of frostbite: 2014 update. Wilderness Environ Med. 2014;25(4 Suppl):S43–54.
57. McMahon JA, Howe A. Cold weather issues in sideline and event management. Curr Sports Med Rep. 2012;11(3):135–41.
58. Gauthier J, Morris-Janzen D, Poole A. Iloprost for the treatment of frostbite: a scoping review. Int J Circumpolar Health. 2023;82(1):2189552.
59. Gatterer H, Dunnwald T, Turner R, Csapo R, Schobersberger W, Burtscher M, et al. Practicing sport in cold environments: practical recommendations to improve sport performance and reduce negative health outcomes. Int J Environ Res Public Health. 2021;18(18)
60. Rauch S, Strapazzon G, Brugger H. On-site medical management of avalanche victims-a narrative review. Int J Environ Res Public Health. 2021;18(19)
61. Zweifel B, Procter E, Techel F, Strapazzon G, Boutellier R. Risk of avalanche involvement in winter backcountry recreation: the advantage of small groups. Wilderness Environ Med. 2016;27(2):203–10.
62. Strapazzon G, Schweizer J, Chiambretti I, Brodmann Maeder M, Brugger H, Zafren K. Effects of climate change on avalanche accidents and survival. Front Physiol. 2021;12:639433.
63. Strapazzon G, Paal P, Schweizer J, Falk M, Reuter B, Schenk K, et al. Effects of snow properties on humans breathing into an artificial air pocket – an experimental field study. Sci Rep. 2017;7(1):17675.
64. Grasegger K, Strapazzon G, Procter E, Brugger H, Soteras I. Avalanche survival after rescue with the RECCO rescue system: a case report. Wilderness Environ Med. 2016;27(2):282–6.
65. Strapazzon G, Beikircher W, Procter E, Brugger H. Electrical heart activity recorded during prolonged avalanche burial. Circulation. 2012;125(4):646–7.
66. Soar J, Berg KM, Andersen LW, Bottiger BW, Cacciola S, Callaway CW, et al. Adult advanced life support: 2020 international consensus on cardiopulmonary resuscitation and emergency cardiovascular care science with treatment recommendations. Resuscitation. 2020;156:A80–A119.
67. Gruber E, Oberhammer R, Balkenhol K, Strapazzon G, Procter E, Brugger H, et al. Basic life support trained nurses ventilate more efficiently with laryngeal mask supreme than with facemask or laryngeal tube suction-disposable--a prospective, randomized clinical trial. Resuscitation. 2014;85(4):499–502.
68. Pasquier M, Hugli O, Paal P, Darocha T, Blancher M, Husby P, et al. Hypothermia outcome prediction after extracorporeal life support for hypothermic cardiac arrest patients: the HOPE score. Resuscitation. 2018;126:58–64.
69. Bonini M, Silvers W. Exercise-induced bronchoconstriction: background, prevalence, and sport considerations. Immunol Allergy Clin N Am. 2018;38(2):205–14.
70. Castellani JW, Young AJ, Ducharme MB, Giesbrecht GG, Glickman E, Sallis RE, et al. American College of Sports Medicine position stand: prevention of cold injuries during exercise. Med Sci Sports Exerc. 2006;38(11):2012–29.

It's Sunny and Bright: UV Exposure

4

Ankur Verma, Joey Brutzkus, Briana Santiago, Hanna Mielke-Maday, and Marcos Henriquez Corporan

4.1 Introduction to Your Specific Environment and Environmental Challenges

Sun exposure management is an issue across athletes of all sports and ages. One study looking at NCAA athletes at Duke and Stanford found that the average athlete spent 4 h per day, 10 months of the year training outdoors, with 84% of these athletes experiencing a sunburn in the last year [1]. Other studies have consistently shown that higher levels of physical activity are correlated with greater skin cancer risks [2].

Ultraviolet rays from the sun include UVA, UVB, and UVC with UVA rays penetrating the dermis and causing indirect DNA damage, leading to wrinkles and some cancers. UVB rays penetrate to the epidermis and cause sunburns, DNA damage, and skin cancers [3]. The World Health Organization UV Index helps classify the risk level with UVI of 1–2 classified as "low," 3–5 as "moderate," 6–7 as "high," 8–10 as "very high," and 11+ as "extreme." UV index can be a good measure for athletes to gauge how long they may be outside and what protective barriers to wear or apply.

UV exposure varies based on sport, time of year, time of day, latitude, altitude, and weather conditions. According to WHO, UV index is highest during the summer months during the 4-h period midday when sun is shining most directly. UVI are greater closer to the equator and at higher altitudes. It is estimated that UV radiation increases by 10% with every 1000 m increase in altitude. Weather and surface conditions also play a role with grass, soil, and water reflecting 10% of UV radiation, sand about 15, sea foam 25%, and snow 80%. Thick cloud cover will reduce exposure, but light clouds do not decrease but simply scatter, UV rays [4].

Several barriers exist that affect athlete's adherence to skin protection strategies including forgetting to apply sunscreen, concerns regarding sunscreen discomfort, performance changes due to slippery feel, inconvenience, application time, eye irritation, evaporative cooling, and acne concerns. UV blocking clothing can additionally feel restrictive or hot [2]. Of note, no differences in skin temperature, sweat rates, or sweat evaporation rates were seen with sunscreen application [5].

A. Verma (✉) · J. Brutzkus · B. Santiago
H. Mielke-Maday · M. H. Corporan
University of Illinois Chicago, Chicago, IL, USA
e-mail: averma28@uic.edu; Joseph.brutzkus@nm.org; Briana_N_Santiago@rush.edu; hanna_mielke-maday@rush.edu; MarcosRafael_HenriquezCorporan@rush.edu

M. R. Hutchinson, S. Rocha Piedade (eds.), *Sports Medicine: It's Environmental*,
https://doi.org/10.1007/978-3-031-84493-5_4

When athletes do wear sunscreen, many do not reapply. One study found that in collegiate golfers and tennis players, 48% applied sunscreen, but only 4.8% reapplied within 2 h [6].

This chapter aims to discuss sun exposure risks and protection strategies in athletes.

4.2 Case Anecdote/Challenge That Introduces Your Unique Environment to the Special Challenges Faced on the Sideline or Competition in Sports Medicine

A 12-year-old middle school baseball player wants to get one last practice before heading to the Little League World Series shown in Fig. 4.1. It's August and 85 degrees and sunny. He's running late to meet his friends at the field and quickly grabs his bag before heading to the field. He wears his usual uniform, including a short-sleeved shirt, knickers, knee-high socks, and baseball cap. He plays out in the field for 2 h before returning home. While showering he notices painful, red skin encompassing the full circumference of his elbows and forearms, as well as over the anterior aspect of his neck.

Fig. 4.1 Middle school baseball player in typical style baseball uniform and gear

A. What sun-protective strategies could he have implemented prior to this practice that could have prevented the skin burns on his arms?
B. What treatment strategies can he pursue at this time to help with symptom management?

The next day he and his parents notice a couple, small- to medium-sized blisters starting to develop over both of his forearms. He doesn't have any fever or chills and there is no drainage from the blisters. He calls his coach and athletic trainer because his first game is tomorrow, and he wants to know if this will keep him from playing in the tournament.

C. What should the coach and athletic trainer recommend to this athlete? Should he be cleared to play? If he is cleared, should there be any specific precautions that the athlete should take?

He is ultimately cleared to play but is kept under strict sun-protective precautions. He plays in 3 games over the next 3 days. The weather has been a little warmer than usual for this time of year, the sun has been out, and his games have been scheduled between 11:00 AM and 2:00 PM. During his fourth game, he presents to his athletic trainer complaining of fatigue and headache, and he can feel his heart beating in his chest, which is unusual for him. The athletic trainer takes his vitals which are found to be heart rate of 115, blood pressure of 105/75, oxygen saturation of 100%, and temperature of 100.3F. The athletic trainer checks the athlete's skin and notices extension of his forearm blisters.

D. What is your differential diagnosis for this athlete?
E. Is this athlete cleared to return to play?
F. What are the next steps in the management and treatment of this athlete?

4.3 Epidemiology of Injury/Illness Risk (Incidence/Prevalence) for Your Specific Environment Both in General and Within Sports Medicine

Sunburns are more common in athletes than in the general population due to increased exposure. Multiple studies have shown increased burn risk, sun exposure, and cancer rates in athletes.

Beginning at the high school level, athletes have higher rates of burns. One study surveyed 21,894 US high school athletes across the nation from 2015 to 2017 and found that sunburn prevalence was about 56.6% during the study time, but in physically active students, the rate was 88.3%. Odds ratios were greatly increased in those who practiced 5+ days a week or those in 3+ sports. There were gender differences, with males at a higher odds ratio to females (2.52 to 1.65, respectively) [7].

Studies have also shown that a higher level of physical activity is associated with a greater risk of melanoma with a hazard ratio of 1.27 [2]. With increased risk of squamous cell carcinoma after 70,000 cumulative hours of sun exposure, athletes who spend many hours outside daily can quickly rack up hours toward this threshold. Basal cell carcinoma risk increases after 2.4 lifetime sunburns [2].

Athletes may find that they burn faster on hot days or with harder workouts as it has been found that sweat significantly decreases the minimal erythema dose. The minimal erythema dose is the lowest UV exposure dose needed to produce erythema. This is thought to be due to sweat decreasing the reflection and dispersion of the UV rays. Studies have shown after 15 min of jogging the minimal erythema dose was decreased by a mean of 17.2% [8].

Water sports can be particularly challenging for sun protection given the uniforms worn in water, UV refraction, sunscreen washing off, and difficulty of reapplication. One study looked at 246 elite water sport athletes ages 16–30 and found season sunburn rate of 76.7%, with 22.6% never using sunscreen [9].

In snow sports, while more body surface area is typically covered by protective clothing, the exposed areas are more at risk. UV index can be 20–30% higher due to altitude as well as reflection off of snow [2]. Skiers have been found to burn after 6 min without sunscreen, and prior research in New Zealand found that 48% of skiers or snowboarders reported getting burned [10].

Although all athletes need sun protection to reduce skin cancer and skin damage risks, skin type also plays a role into burn prevalence, with 34–66% of black skin respondents reporting no history of prior sunburn [11]. Overall, skin cancer risk may be lower for those with darker skin than those with light skin, but skin cancers that develop in darker-skinned individuals tend to have higher morbidity and mortality. Therefore, it is prudent to encourage all athletes to use photoprotection regardless of propensity to burn [12].

4.4 Unique Injuries/Illnesses/Patterns Within Your Specific Environment

The sports environment has a multifactorial effect on ultraviolet radiation. Factors such as altitude and ground reflection are crucial elements. Studies have shown that ultraviolet radiation can increase up to 4% per 1000 ft. climbed due to decreased atmospheric absorption; surfaces such as grass and rocks can reflect between 2% and 5% and water 10% and 25%; however, snow has a notoriously high rate above 90% [13].

Water and snow sports are known to increase the risk of acute UVB-induced corneal injury, also known as photokeratitis, due to the UB ground-reflective properties of snow and water, while chronic UVA exposure increases the risk of macular degeneration [14]. Snow sports carry

Table 4.1 Summary of sports environments and the impact on sun and ultraviolet light exposure

Sports modality	UV-induced injury	Anatomical location	UV ground-reflective properties
Outdoor land athletes	Elevated nevi Atypical nevi Solar lentigines	Shoulder Buttock	2–5% on grass and rocks
Outdoor aquatic athletes	Basal cell carcinoma melanoma Squamous cell carcinoma	Upper extremities	10–25% on water
Snow athletes	Photokeratitis	Cornea	90% on snow

additional risk due to ground elevation as noted by skiers and snowboarders. At 11154.9 ft. of altitude, skiers without skin protection can experience sunburns after only 6 min [15].

Aquatic athletes with chronic sunburn injuries are at high risk for carcinogenic lesions. Among those malignancies, BCC is the most common skin cancer at 71%, followed by melanoma at 18%, and SCC at 10%. The head was the most common location followed by upper extremities [16].

Marathon runners tend to have elevated nevi, atypical nevi, and solar lentigines, primarily seen in the shoulder and buttocks [17]. See Table 4.1 for summary of sports environments and the impact on sun and UV light exposure.

4.5 Management of Injuries/Illness at Venue (Within Environment) and Immediately After While Still with Team But Before Transition Home

Acute sunburns are common in outdoor athletes and primarily consist of symptom management at home, rarely requiring escalation of care; see Fig. 4.2. However, extensive injuries require immediate intervention.

While in the venue, athletes should avoid further sun exposure and protect their skin by using sunscreen and sun-protective clothing to prevent injury progression. Pain is one of the most predominant symptoms and taking over-the-counter NSAIDs can help reduce pain [18]. Nonpharmacological treatments consist of optimizing hydration; cool compresses, aloe vera, or lotion can soothe the skin, while cool colloidal oatmeal baths or calamine lotion can also provide relief [19]. Local anesthetic creams and topical and oral corticosteroids are contraindicated, as they do not benefit sunburn treatment [20].

Sunburns resulting in blisters should be gently washed with soap and water. If blisters rupture, athletes should cover them with moist gauze, and intact blisters should not be opened [18]. Severe sunburns leading to extensive blistering and potential fluid and electrolyte loss might require medical intervention, particularly if more than 20% of the body surface area is affected [21]. Treatment might involve fluid resuscitation using the Parkland formula (4 mL × %TBSA × weight in kg) or the modified Brooke formula (2 mL × %TBSA × weight in kg), administering half of the calculated fluid within the first 8 h post-injury and the rest over the next 16 h [22]. The resuscitation should begin immediately after the injury, irrespective of the emergency department's arrival time. Severe cases might necessitate treatment at a burn center [18].

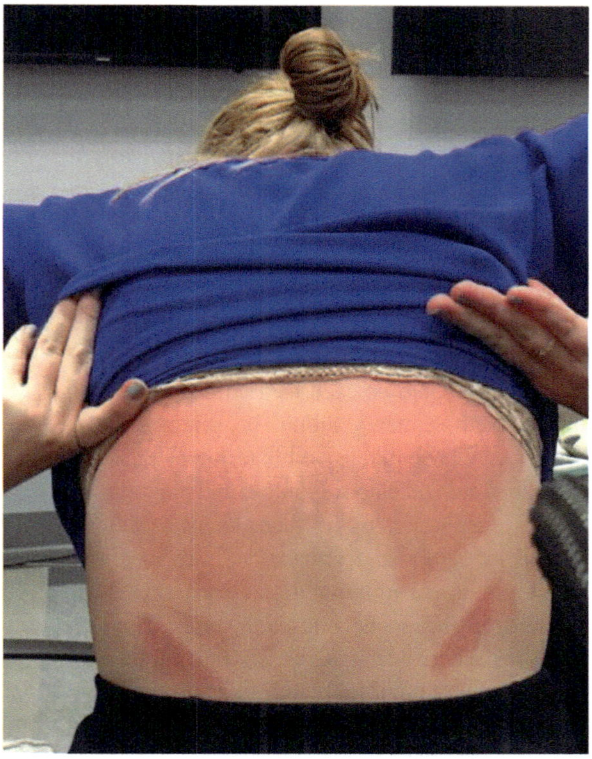

Fig. 4.2 Acute sunburn in an outdoor cyclist after a long ride. Notice the pattern of the sunburn which represents exposed skin versus skin covered by clothing and gear

4.6 Extraction Challenges for Your Specific Environment

Extraction from sun exposure depends largely on the degree of sunburn. Athletes with first-degree burns can return to play with caution. Hydration should be emphasized. If possible, athletes should avoid sunlight during peak UV radiation times between 10:00 AM and 4:00 PM and seek shade when able as shown in Fig. 4.3. Broad-spectrum, water-resistant sunscreen with an SPF of 30 or more blocking UVA and UVB rays should be applied at least 15 min before sun exposure. Sunscreen should be applied at least every 2 and after swimming, profusely sweating, or toweling off. Athletes should wear sun-protective clothing which includes darker clothing with UPF (ultraviolet protection factor) of 40–50, long sleeve shirts with higher necklines, long pants, sunglasses with UV-protective lenses, and wide-brimmed hats. Extra caution should be applied for those participating in sun-reflective environments such as snow, water, sand, and athletes closer to the equator and high altitudes, which all place athletes at higher risk.

Fig. 4.3 Track and field athletes seeking shade under a tent between events to limit sun exposure during the day-long competition

First-degree burns should improve within 72 h. Symptoms such a skin redness, pain, pus-like discharge, or temperature of 100.4 °F or 38 °C are signs that the athlete may have a worsening degree of burn or infection [23]. In this case, the athlete should be removed from play until evaluated by a healthcare provider.

Athletes with second-degree burns should avoid outdoor activities, if possible, especially if the required uniform does not allow the area of skin burn to be covered. If the area of skin burn is easily covered and the athlete wishes to return to play, they must implement the sun-protective strategies mentioned above. If the burn area worsens or does not improve, complete removal from sport should be exercised until skin healing is observed. Athletes should be removed from play if they have extensive blistering as this may result in fluid and electrolyte loss and may require fluid resuscitation [24]. Additionally, athletes should be removed from play if they begin to develop extreme inflammation of the skin, itching, blisters, headache, nausea, fever, or dehydration [25]. These are signs of polymorphous light eruption, more commonly known as "sun poisoning," which can occur within a few hours of being in the sun and can last anywhere from two days to weeks [25].

Any athlete with third-degree burns should be removed from play. Third-degree burns will usually require hospitalization. Resuscitation with a crystalloid solution may be necessary if the total body surface area (TBSA) affected exceeds 20% in adults and children [24]. In this instance, administration of the crystalloids will occur in a stepwise fashion, giving half of the crystalloid solution within the first 8 h after the burn and the remaining half over the following 16 h [24]. Third-degree burns are generally best treated at a burn center and treatment should begin immediately after the injury.

It is important to recognize and consider heat exhaustion in athletes exposed to UV radiation. Symptoms of heat exhaustion include dizziness, heavy sweating, nausea, vomiting, extreme thirst, headache, fatigue, tachycardia, and cool moist skin with piloerection [26]. These athletes should be identified immediately and extracted from play. In addition to the symptoms of heat exhaustion, if the athlete then begins to develop confusion or loss of consciousness, they have progressed to heat stroke [26]. In the event of heat stroke, the athlete should also be immediately removed from play, cooled, and sent to the hospital for further evaluation and management. The topic of heat stroke will be discussed in further detail in another chapter.

4.7 Prevention Plans and Programs Including Education and Intervention to Reduce or Avoid Future Injury/Illness in Your Environment

4.7.1 Sunscreen

Athletes should choose a broad-spectrum sunscreen that blocks ultraviolet A (UVA) and ultraviolet B (UVB) radiation [27]. Athletes should choose a water-resistant sunscreen if sweating or involved in swimming. A sun protection factor (SPF) of 30 or greater is recommended and should be applied at least every 2 h [28]. Higher SPF of 50 and up to 100 may be beneficial for water sports and snow sports when the UV index is high or very high [29]. Athletes may need to apply more frequently if toweling off, swimming, or with excessive perspiration [28]. The application of sunscreen should be done at least 15 min prior to going outdoors [28]. This not only allows proper absorption of the sunscreen but can also help prevent sunscreen dripping into and irritating the eyes and can help prevent feelings of the hands being slippery after application, which can improve compliance.

4.7.2 Clothing

Clothing can also protect the athlete from harmful UV radiation by absorbing or blocking radiation [30]. Clothing should be breathable with tightly woven materials incorporating ultraviolet protection factor (UPF) [31]. UPF describes what fraction of the sun's ultraviolet radiation can penetrate the fabric. The higher the UPF, the better. For example, a fabric with UPF rating of 50 permits 1/50th of the sun's ultraviolet radiation to reach the skin [32]. A UPF of 40–50 provides excellent UV protection by transmitting less than 2.6% of UV radiation [31]. When compared to lighter color fabric, darker color fabric will better absorb UV radiation, keeping it from reaching the skin's surface [33]. The more skin that can be covered, the better. This includes higher necklines, long sleeves, rash guards for surfers, pants, hats, and sunglasses with ultraviolet-protective lenses [26]. Some of these options will not be possible given certain uniform requirements, excessive heat, or restriction of movement that athletes may experience with extensive skin coverage.

4.7.3 Time of Activity

It is best to avoid sun exposure between 10:00 AM and 4:00 PM when UV radiation is the highest [34]. When possible, practices and games should be scheduled in the morning or late afternoons. When unable to avoid peak UV radiation hours, athletes should seek out or create shade with umbrellas, tents, or shaded seating when not playing.

4.7.4 Medications

It is important to consider medications that can make the skin more sensitive to sunburn. Some medications that have been found to make the skin more sensitive to burns include nonsteroidal anti-inflammatories, tetracyclines, ciprofloxacin, furosemide, hydrochlorothiazide, phenothiazines, amiodarone, antifungals, and retinols [35]. Athletes requiring these medications should take extra precautions when participating in outdoor activity.

Multiple studies have measured UV exposure and cancer rates among athletes and many initiatives are promoting skin and sun education for athletes. SUNSPORT (Stanford University Network for Sun Protection, Outreach, Research and Teamwork), Go Sun Smart, and Clever in Sun and Shade Programs are just a few of these initiatives [2]. Improvement in athlete awareness of UV radiation and promotion of consistent application of sun-protective strategies continue to be explored and developed. One strategy includes UV detection stickers that are visibly displayed to athletes and electronic or phone applications to track personal UV exposure [2]. See Table 4.2 for a summary of sun exposure prevention strategies.

Table 4.2 Summary of sun exposure prevention strategies

Sunscreen application	Clothing	Time of day	Medication review
Use broad-spectrum, water-resistant products that block UVA and UVB radiation	The more skin that can be covered, the better. Long sleeve shirts, shirts with high neck lines, sunglasses with UV-protective lenses, and wide-brimmed hats are preferred	Avoid sun exposure between 10:00 am and 4:00 pm	Review medications that can make the skin more sensitive to sunburns (NSAIDs, quinolone, tetracycline antibiotics, diuretics, HCTZ, psoralens, and phenothiazines, amiodarone, antifungals, retinols)
Reapply sunscreen every 80 min. Reapply after swimming, perspiring, or toweling off	Incorporate clothing with UPF (ultraviolet protection factor). The higher the UPF, the greater the protection	Seek out or create shade with a tent or umbrella to avoid peak UV time	
SPF should be at least 30 of greater. Higher SPF does not mean you can apply less	Fabric color matters. Darker color fabrics will absorb more UV rays so fewer UV rays reach the skin		
Apply sunscreen at least 15 min prior to activity or going outdoor to allow for full skin absorption and help prevent slippery hands or eye irritation			

There have also been campaigns aimed to educate athletes, parents, coaches, trainers, referees, and spectators on the importance of skin protection and offer approaches to prevent skin damage from UV radiation. These campaigns also seek to modify policies surrounding uniforms and eye protection to allow for more flexibility. There has also been a push to have visual cues of sun safety in locker rooms along with easily accessible sunscreen [2]. Prior to large, outdoor events, organizers could send sun safety reminders to all in attendance and provide reminders throughout the event via stadium speakers and jumbotron [2]. Select universities and organizations have started to implement skin screening at outdoor sporting events.

4.8 Take-Home Pearls Bulleted as Summary

- Both burn and skin cancer rates are higher in athletes due to increased outdoor sun exposure.
- Conditions that impact burn risk include cloud cover, altitude, latitude, time of day, reflection from snow or water, and even the amount of sweat on the skin.
- Athletes should have sunscreen available in the locker room and on site for reapplication, and application should be encouraged by coaches and staff, which has been proven to increase use.
- Sunscreen should be applied 15 min before outdoor activity, reapplied every 80 min, and should be at least SPF 30.
- If possible, dark clothing, sunglasses, hats, and other skin protection should be worn.
- Athletes should be conscious of additional medications they might be on which increase photosensitivity.
- Athletes should be removed from play if they begin to develop extreme inflammation of the skin, itching, blisters, headache, nausea, fever, and dehydration.

References

1. Wysong A, Gladstone H, Kim D, Lingala B, Copeland J, Tang JY. Sunscreen use in NCAA collegiate athletes: identifying targets for intervention and barriers to use. Prev Med. 2012;55(5):493–6. https://doi.org/10.1016/j.ypmed.2012.08.020.
2. Gilaberte Y, Trullàs C, Granger C, de Troya-Martín M. Photoprotection in outdoor sports: a review of the literature and recommendations to reduce risk among athletes. Dermatol Ther. 2022;12(2):329–43. https://doi.org/10.1007/s13555-021-00671-0.
3. UV rays: types of ultraviolet rays; UVA, UVB, UVC and cancer [Internet]. City of Hope (2022) [cited 2024 Mar 28]. Available from: https://www.cancercenter.com/cancer-types/skin-cancer/risk-factors/uv-rays
4. Radiation: Ultraviolet (UV) radiation [Internet]. World Health Organization; [cited 2024 Mar 28]. Available from: https://www.who.int/news-room/questions-and-answers/item/radiation-ultraviolet-(uv)
5. Ou-Yang H, Meyer K, Houser T, Grove G. Sunscreen formulations do not interfere with sweat cooling during exercise. Int J Cosmet Sci. 2017;40(1):87–92. https://doi.org/10.1111/ics.12440.
6. Sellyn GE, Millan SH, Harmon KG, Karpinos A. A survey of sun safety habits among division I national collegiate athletic association golfers and tennis players. Clin Dermatol. 2023;41(6):729–34. https://doi.org/10.1016/j.clindermatol.2023.09.015.
7. Wei G, Farooq J, Castelo-soccio L, Mhaskar R. Correlates between physical activity and sunburn prevalence among a nationally representative sample of US High School students, 2015–2017. J Phys Act Health. 2021;18(9):1113–9. https://doi.org/10.1123/jpah.2020-0711.
8. Jinna S, Adams BB. Ultraviolet radiation and the athlete: risk, sun safety, and barriers to implementation of protective strategies. Sports Med. 2013;43(7):531–7. https://doi.org/10.1007/s40279-013-0021-5.
9. De Castro-Maqueda G, Gutierrez-Manzanedo JV, Lagares-Franco C, de Troya-Martin M. Sun exposure during water sports: do elite athletes adequately protect their skin against skin cancer? Int J Environ Res Public Health. 2021;18(2):800. https://doi.org/10.3390/ijerph18020800.
10. Price J, Ness A, Leary S, Kennedy C. Sun-safety behaviors of skiers and snowboarders on the South Island of New Zealand. J Cosmet Dermatol. 2006;5(1):39–47. https://doi.org/10.1111/j.1473-2165.2006.00221.x.
11. Diffey BL, Fajuyigbe D, Wright CY. Sunburn and sun protection in Black Skin. Int J Dermatol. 2019;58(9):1053–5. https://doi.org/10.1111/ijd.14402.
12. Goon P, Banfield C, Bello O, Levell NJ. Skin cancers in skin types IV–VI: does the Fitzpatrick scale give a false sense of security? Skin Health Dis. 2021;1(3):e40. https://doi.org/10.1002/ski2.40.

13. Willmann G. Ultraviolet keratitis: from the pathophysiological basis to prevention and clinical management. High altitude medicine & biology [Internet] 2015 [cited 2019 Dec 29];16(4):277–82.
14. Rodriguez JO, Lavina AM, Agarwal A. Prevention and treatment of common eye injuries in sports [Internet]. American Academy of Family Physicians; 2003 [cited 2024 Mar 17]. Available from: https://www.aafp.org/pubs/afp/issues/2003/0401/p1481.html
15. Gilaberte Y, Trullàs C, Granger C, de Troya-Martín M. Photoprotection in outdoor sports: a review of the literature and recommendations to reduce risk among athletes. Dermatol Ther (Heidelb) [Internet]. 2022 [cited 2024 Mar 18];12(2):329–43. Available from: https://doi.org/10.1007/s13555-021-00671-0
16. Bennett HG, Dahl LA, Furness J, Kemp-Smith K, Climstein M. Skin cancer and sun protective behaviours in water-based sports: a scoping review. Photodermatol Photoimmunol Photomed [Internet]. 2022 [cited 2024 Mar 18];38(3):197–214. Available from: https://pubmed.ncbi.nlm.nih.gov/34582598/
17. Snyder A, Valdebran M, Terrero D, Amber KT, Kelly KM. Solar ultraviolet exposure in individuals who perform outdoor sport activities. Sport Med Open. 2020;6(1)
18. UpToDate [Internet]. Uptodate.com. [cited 2024 Mar 28]. Available from: https://www.uptodate.com/contents/skin-burns-beyond-the-basics?topicRef=2730&source=see_link
19. Guerra KC, Crane JS. Sunburn. StatPearls Publishing; 2023.
20. Han A, Maibach HI. Management of acute sunburn. Am J Clin Dermatol [Internet]. 2004;5(1):39–47. Available from: https://doi.org/10.2165/00128071-200405010-00006
21. Żwierełło W, Piorun K, Skórka-Majewicz M, Maruszewska A, Antoniewski J, Gutowska I. Burns: classification, pathophysiology, and treatment: a review. Int J Mol Sci [Internet]. 2023 [cited 2024 Mar 28];24(4):3749. Available from: https://doi.org/10.3390/ijms24043749
22. Mehta M, Tudor GJ. Parkland Formula. StatPearls Publishing; 2023.
23. Burns: Medlineplus medical encyclopedia [Internet]. U.S. National Library of Medicine; [cited 2024 Mar 27]. Available from: https://medlineplus.gov/ency/article/000030.htm
24. Guerra KC. Sunburn [Internet]. U.S. National Library of Medicine; 2023 [cited 2024 Mar 03]. Available from: https://www.ncbi.nlm.nih.gov/books/NBK534837/#:~:text=People%20suffering%20from%20severe%20sunburns,)%20exceeds%2020%25%20in%20adults
25. Oakley AM, Ramsey ML. Polymorphic Light Eruption [Internet]. NHS; [cited 2024 Mar 21]. Available from: https://www.nhs.uk/conditions/polymorphic-light-eruption/
26. Heat stress – heat related illness [Internet]. Ctr Dis Control Prev; 2022 [cited 2024 Mar 25]. Available from: https://www.cdc.gov/niosh/topics/heatstress/heatrelillness.html#:~:text=Heat%20Stroke,-Heat%20stroke%20is&text=It%20occurs%20when%20the%20body,within%2010%20to%2015%20minutes
27. Adams BB. Dermatologic disorders of the athlete. Sports Med. 2002;32(5):309–21. https://doi.org/10.2165/00007256-200232050-00003.
28. Diffey BL. When should sunscreen be reapplied? J Am Acad Dermatol. 2001;45(6):882–5. https://doi.org/10.1067/mjd.2001.117385.
29. Williams JD, Maitra P, Atillasoy E, Wu M-M, Farberg AS, Rigel DS. SPF 100+ sunscreen is more protective against sunburn than SPF 50+ in actual use: results of a randomized, double-blind, split-face, natural sunlight exposure clinical trial. J Am Acad Dermatol. 2018;78(5):902. https://doi.org/10.1016/j.jaad.2017.12.062.
30. Gambichler T, Altmeyer P, Hoffmann K. Role of clothes in sun protection. Cancers of the. Skin. 2002:15–25. https://doi.org/10.1007/978-3-642-59410-6_3.
31. Kliniec K, Tota M, Zalesińska A, Łyko M, Jankowska-Konsur A. Skin cancer risk, sun-protection knowledge and behavior in athletes—A narrative review. Cancers. 2023;15(13):3281. https://doi.org/10.3390/cancers15133281.
32. Sun Protective Clothing [Internet]. 2021 [cited 2024 Mar 27]. Available from: https://www.skincancer.org/skin-cancer-prevention/sun-protection/sun-protective-clothing/#:~:text=Ultraviolet%20Protection%20Factor%20(UPF)%20indicates,reducing%20your%20exposure%20risk%20significantly
33. Sunscreens and photocarcinogenesis. Clinical Guide to Sunscreens and Photoprotection 2008 Dec 22;105–16. https://doi.org/10.3109/9781420080858-10.
34. Sun Exposure [Internet]. Centers for Disease Control and Prevention; 2018 [cited 2024 Mar 27]. Available from: https://www.cdc.gov/niosh/topics/sunexposure/default.html
35. Photosensitivity report – medications [Internet]. 2021 [cited 2024 Mar 14]. Available from: https://www.skincancer.org/risk-factors/photosensitivity/medications/

It's Stormy

5

Julio Castillo Tafur, Nirav Mungalpara, and Mark R. Hutchinson

5.1 Introduction

Stormy conditions present unique challenges in the field of sports medicine. Heavy rain, thunderstorms, lightning, hurricanes, or even tornadoes can drastically affect the safety and well-being of athletes, spectators, and medical teams. Proper planning and disaster preparedness are essential to mitigate the risks associated with such conditions, ensuring that events can be managed safely and effectively.

Understanding and preparing for stormy conditions is critical due to their unpredictable nature and potential for severe impact. For instance, lightning injuries, while rare, can be life-threatening and require immediate medical attention. Hurricanes and tornadoes also pose significant threats, not only due to their destructive power but also because of the secondary effects, such as flooding and structural damage. These events can disrupt sporting activities and necessitate rapid and effective response strategies.

This chapter aims to equip sports medicine professionals with the knowledge and tools necessary to navigate the challenges posed by stormy weather. We will explore real-world case studies to highlight specific scenarios, delve into the epidemiology of weather-related injuries, and discuss the unique types of injuries that can occur. Additionally, we will provide management strategies for treating injuries on-site, address the challenges of extracting individuals from hazardous environments, and outline preventive measures to reduce future risks. By the end of this chapter, readers will have a comprehensive yet practical reference to handle stormy conditions in sports settings.

5.2 Case Study

You are asked to serve as the event physician for an outdoor track and field competition. Pre-event planning included an assessment of weather. Temperature was predicted to be about 27–29 °C (80–85 °F), but thunderstorms and lightning were also predicted. At the time of the competition, lightning is seen in the distance. Who makes the call to evacuate the field of play? The head coach? Each individual coach of their unique event? The team physician? The venue manager? The event physician?

The scenario actually occurred with no definition of "who was in charge." Some athletes wanted to only listen to their coach. The head coach wanted to defer to his team physician (who he knew would take the risk). The venue manager was distracted as they were in charge of three

J. C. Tafur (✉) · N. Mungalpara
University of Illinois at Chicago, Department of Orthopaedics, Chicago, IL, USA
e-mail: jcastil@uic.edu

M. R. Hutchinson
UIC Sports Medicine Center, Department of Orthopaedics, University of Illinois at Chicago, Chicago, IL, USA

M. R. Hutchinson, S. Rocha Piedade (eds.), *Sports Medicine: It's Environmental*,
https://doi.org/10.1007/978-3-031-84493-5_5

venues simultaneously. Ultimately, the event physician assumed responsibility but with some coaches and athletes ignoring the evacuation order as they disagreed with "who was in charge." No injury occurred, but the next time an event occurred, a clear ladder of responsibility was established to assure everyone would comply with the evacuation order as needed (Table 5.1).

The following represent other real-world weather-related events that impacted sport:

Table 5.1 Trauma evaluation (ABCDE approach)

A	Airway
B	Breathing
C	Circulation
D	Deformity
E	Exposure

Condition	Event (year)	Impact reported
Lightning strike	Soccer match in Johannesburg (1998) [1, 2]	Several players and spectators were injured when lightning struck during the match
	Soccer match in Democratic Republic of Congo (1998) [3]	A lightning strike killed all 11 members of one team during a match in Bena Tshadi, Eastern Kasai Province. The other team was left completely unharmed, leading to suspicions of witchcraft. More than 30 spectators were also injured
	Under-20 soccer match in Sao Paulo (2019) [4]	Brazilian footballer Henrique, playing for Agua Santa, was struck by lightning during an under-20 match against Atletico Mineiro. The match was halted due to a thunderstorm and torrential downpour. Henrique made a full recovery after being struck
	Soccer match in Indonesia (2024) [2]	Indonesian footballer Septian Raharja, 35, was struck by lightning during a friendly match at Siliwangi Stadium, Bandung. He suffered severe burns and succumbed to his injuries at the hospital. His cleats caught fire, and his jersey was torn due to the intensity of the strike
Typhoon	Rugby World Cup in Japan (2019) [5, 6]	Several matches were canceled due to Typhoon Hagibis, leading to logistical challenges and safety concerns for athletes and fans
Heavy rain/ flooding	Brazilian Grand Prix (2016) [7, 8]	Heavy rainfall caused multiple crashes, leading to the race being red-flagged twice. The flooding also disrupted logistics and event schedules
Hurricane	Hurricane Harvey (2017) [9]	1. No college football LSU Tigers vs. BYU opener relocated from NRG Stadium, Houston UT San Antonio canceled Football Fan Day The Houston Cougars moved practice to Austin 2. NFL (National Football League) The Houston Texans and Dallas Cowboys considered relocating the preseason finale to Dallas 3. MLB (Major League Baseball) The Houston Astros considered moving or postponing the series against the Texas Rangers
	Hurricane Dorian (2019) [10]	1. College football The game between the FSU Seminoles and BSU Broncos was moved from Jacksonville to Tallahassee, and kickoff time for the game was also changed 2. NFL (National Football League) The Jacksonville Jaguars made preparations for potential impacts on their home game against the Kansas City Chiefs 3. MLB (Major League Baseball) The Tampa Bay Rays prepared contingency plans for their series against the Baltimore Orioles Plans included potential postponements or relocation
	Hurricane Andrew (1992) [11]	1. National Football League (NFL) *The Miami Dolphins had to move their preseason games and practices*
Floods	Calgary floods (2013) [12]	1. *National Hockey League (NHL)*: The Calgary Flames were forced to relocate their training camp and preseason activities as the Scotiabank Saddledome suffered significant damage
Tornadoes	Joplin tornado (2011) [13]	1. *Minor League Baseball*: The Joplin Blasters had to cancel games and adjust their schedule as the community recovered from the devastating tornado

5.3 Epidemiology

Lightning injuries are rare but potentially life-threatening. It is estimated 444 deaths in the USA were caused by lightning from 2006 through 2021. People are most at risk in the southeastern states, with Florida being considered the "the lightning capital" of the USA. Most of these injuries occur while performing outdoor activities in the summer months, particularly in the month of July. Despite roughly at least 40 million lightning strikes per year, the chance of being hit by one is less than one in million for any given year. Your risk of being struck by lightning increases significantly with outdoor activities and working [1] (Table 5.2).

Hurricanes, also called tropical cyclones, are low-pressure systems that have organized thunderstorms without forming a front. These storms are common in tropical and subtropical regions. Hurricanes are defined as storms whose sustained winds are at least 74 miles per hour (mph) [2]. Hurricanes are further classified based on their maximum sustained winds using the Saffir-Simpson Hurricane Scale. Hurricanes are rated from 1 to 5, with higher categories representing higher wind speeds and greater potential for property damage. Major hurricanes are often categories 3 and above. While the Saffir-Simpson scale classifies hurricanes on potential for damages, it does not consider other possibly associated phenomena such as storm surges, flooding, or tornadoes [3].

Hurricane season begins on June 1 in the Atlantic and Caribbean. Hurricane season ends roughly around November 30. These storms tend to organize in the Atlantic basin (Atlantic Ocean, Caribbean Sea, and Gulf of Mexico). On average, hurricanes are identified 12 times a year in this region [2]. It goes without saying that hurricanes have tremendous potential for casualties and property damage. The deadliest hurricane in US history, the Galveston hurricane, was responsible for at least 8000 deaths. In contrast, Hurricane Katrina in 2005 was responsible for approximately $108 billions of property damage, making it the costliest hurricane in the US recent history [4].

Table 5.2 Orientation assessment

Person	Able to relate person = 1	Unable to relate person = 0
Place	Able to relate place = 1	Unable to relate person = 0
Time	Able to relate time = 1	Unable to relate time = 0

Commonly reported A&O x 0 to 3

Tornadoes are rare and unpredictable rotating columns of air that extend from a thunderstorm to the ground. Roughly 1200 tornadoes are reported in the USA yearly. Tornadoes have been reported in all fifty states of the USA and tornado season varies by geographically. They can occur at any time of the year and at any time of the day. In the Southern Plains, tornadoes are most common in the months of May and June. On the Gulf Coast, tornadoes are more likely during the spring season. In the Upper Midwest and Northern Plains, tornado season usually occurs between the months of June and July [5]. The deadliest tornado in recent US history occurred in 1925 in the tristate area of Missouri, Illinois, and Indiana with approximately 695 deaths [6].

5.4 Unique Injuries

5.4.1 Lightning

While athletic injury is the priority of the sports medicine physician, knowledge of the possible injuries that could be sustained to the spectators is imperative. Interestingly, lightning injuries differ from high-voltage injuries. The particular injuries seen in lightning victims are the result of electrical conduction rather than electrical burns. The different tissues and organ systems affected by lightning depend on the relative resistance of tissues [7], with muscles and nerves being the least resistant to lightning conduction pathways [8].

Cardiac injuries associated with lightning injuries deserve special attention. Cardiopulmonary arrest is by far the most severe. It is speculated that cardiac arrest is brought on by respiratory arrest due the effect lightning has on the respiratory centers of the medulla. Interestingly, survivors of cardiac arrest after lightning strike are very unlikely to pass. Some other signs of cardiac injury may be observed in electrocardiograms of lightning victims, and these include ventricular tachycardia, premature ventricular contraction, atrial fibrillation, prolonged QT interval, ventricular fibrillation, and ST changes [8].

The nervous system is also quite susceptible to lightning injuries, and these injuries may have long-lasting effects on patients. Blunt trauma to the brain, cerebral edema, intracranial hemorrhage, or seizures may be observed in lightning injuries. The sonic blast and pressure differential, also known as barotrauma, following a strike may be responsible for traumatic injuries to the brain. The autonomic nervous systems may also be affected leading to movement disorders, edema, complex regional pain syndrome, or other autonomic disorders. At last, post-traumatic stress disorder, mood changes, insomnia, decreased libido, and other psychiatric changes may manifest after a lightning strike [9].

Keraunoparalysis (or Charcot's paralysis) is a transient paralysis often involving the lower extremities with associated loss of sensation. This neurological disorder is specific to lightning victims, and it is often short-lived lasting several hours. This transient paralysis is thought to be due to a catecholamine surge, as it is accompanied by pallor, vasoconstriction, and hypertension. Nonetheless, victims with neurological deficits after a lightning strike should be assumed to have a spinal cord injury until proven otherwise [9].

Cold, clammy, and pulseless extremities may also signal possible muscle injury and compartment syndrome. Patients should be monitored for elevated compartment pressures; however, spontaneous return of pulses and warmth are often the outcome. Myoglobinuria resulting from muscle damage is more common in high-voltage injuries rather than lightning injuries [9].

Injuries to the integumentary systems after lightning strikes present in the form of linear, punctate, feathering, or thermal burns. Linear burns are roughly 1 to 4 cm in diameter and may be found under the breasts, arm, or middle of the chest, as these follow areas of high sweat concentration. Thought to be due to water vaporization on the skin surface, these may be seen immediately or may develop over several hours. Punctate burns are small circular lesions often less than1 cm. These may present at the tips of the toes or feet soles. Classically, lightning injuries are associated with Lichtenberg figures, which are a type of feathering lesion. These are not thermal injuries as the dermis and epidermis are normal and often disappear without residual effects. At last, thermal burns are often associated with earring metallic objects such as zippers or if clothes are ignited by the lightning strike [9].

While not always readily apparent, ophthalmologic injuries may be caused by a lightning strike. Cataracts may ensue in lightning victims, but these may not appear until several years after injury. Other injuries include corneal burn, intraocular hemorrhage, uveitis, chorioretinitis, iridocyclitis, hyphema, or optic nerve injury [10]. First responders should be reminded that dilated pupils are not a reliable sign of brain death in lightning strike victims [10].

The barotrauma caused by a lightning strike may lead to tympanic membrane rupture or sensorineural damage. Ossicular disruption may also be observed. Deafness may be seen in these patients, which may be temporary or permanent depending on the nature of the injury [11] (Table 5.3).

5.4.2 Hurricanes

Hurricanes have the capacity to rapidly overwhelm major states response teams. Preparation and familiarity with the common injuries seen during these major environmental disasters is of the essence. The injuries that occur during the hurricane are different from those that occur in its aftermath. For instance, the most common sources of injury and death during the storm are drowning from floods and storm surges and blunt injury caused by debris carried by high-speed winds [12]. In contrast, common injuries in the aftermath of hurricanes

Table 5.3 GCS (LOC assessment)

	1	2	3	4	5	6
Eye (E)	Does not open eyes	Opens eyes in response to pain	Opens eyes in response to voice	Opens eyes spontaneously	N/A	N/A
Verbal (V)	Makes no sounds	Incomprehensible sounds	Inappropriate words	Confused and disoriented but able to answer questions	Oriented to time, person, and place; converse normally	N/A
Motor (M)	Makes no movements	Abnormal extension (decerebrate posture)	Abnormal flexion (decorticate posture)	Withdrawal to painful stimuli	Moves to localize pain	Obeys commands

Commonly reported as GCS of 3–15
Created by Graham Teasdale and Bryan Jennett in 1974. Clinical scale to assess level of consciousness

are due to electrocution, cuts and slips, and infections [12]. Floodwaters may expose survivors to respiratory, gastrointestinal, or vector-borne illnesses.

Sports medicine physicians should be aware that half of tornado-related injuries occur during rescue attempts and cleanup. Often, these injuries involved navigating an unfamiliar environment, as roughly one third of the injuries sustained during a tornado in Illinois were due to stepping on nails. Similarly, injuries may involve fire, electrocution, or explosion [13].

Often undiagnosed, carbon monoxide (CO) poisoning has been observed in the aftermath of major disasters such as hurricanes, floods, and winter storms. Particularly, generator use during these disaster scenarios in situations has been associated with fatal and nonfatal cases of CO poisoning [14, 15]. It is worth reminding the sports medicine physician that CO poisoning may be observed in other sporting and recreational activities including mountain climbing, ice hockey, or boating. Hence, the sports medicine physician taking care should keep a high index of suspicion, particularly during natural disasters [16].

Signs and symptoms of CO poisoning, a recent history of CO exposure, and elevated serum carboxyhemoglobin (COHb) levels are often needed to diagnose CO poisoning. Patients often endorse headaches, nausea/vomiting, altered mentation, fatigue, chest pain, shortness of breath, or loss of consciousness [17] .

5.5 Management

In the event of lightning or suspected lightning, postpone all activities if a storm is expected during or before the event. First aid should be provided to lightning victims. Responders should be mindful of their safety prior to assisting the injured. Therefore, evacuation of the injured patient to a safe location is warranted during an active storm. While responders might be at risk of lightning injury if active storms continue, touching a lightning victim is safe and does not carry a risk of electrocution. Of note, it should be noted that this does not apply to patients electrocuted by fallen power lines. Care should be taken to avoid touching these patients until the scene is safe [18].

The mortality of lightning strike victims is less than 10%. In the event of cardiorespiratory arrest, first responders should initiate CPR. Priority should be given to these victims, as those breathing and moving will likely survive. This is often referred to as "reverse triage." An automated external defibrillator (AED) should be used on the unconscious or those in cardiac arrest. It is imperative to follow the directions of AED, as the AED will deliver an electrical shock for those with a shockable rhythm. CPR should be promptly resumed for those who do not present with such cardiac rhythm. Life-saving measures should not be halted, however, in favor of searching for an AED [18]. Once return of spontaneous circulation (ROSC) has been obtained, it

Fig. 5.1 Lightening position. (taken from Davis et al. [19]))

is necessary to secure and maintain ventilation. Failure to do so may precipitate a second cardiac arrest. [19] Therefore, it is necessary for the responding team to be ready to perform CPR should the patient need to be re-resuscitated. The following figure from the American Heart Association may be used as a reference for the treating sports medicine physician (Fig. 5.1).

Lightning burns often require no treatment besides observation given their superficial nature and quick healing potential. While unlikely, emergent fasciotomy should be considered in patients with cold, clammy, and pulseless extremities with no clear sign of resolution and signs of compartment syndrome [9]. If so, prompt transport and coordination with a surgical service should be sought. On the other hand, all lightning victims should seek ophthalmologic consultation and monitoring, as cataracts may develop in this patient population [9, 10]. Simple tympanic ruptures may be treated conservatively, but patients with loss of hearing should seek otolaryngology consultation [11].

CO poisoning is currently treated with either 100% normo-baric oxygen or hyperbaric oxygen therapy if available [8]. Hyperbaric therapy may, in fact, be more effective in improving severe neurological dysfunction as a result of CO poisoning [17]. Pulse oximeters may be used to monitor the patient; however, pulse oximeters cannot be used to diagnose CO poisoning due to its inability to detect COHb [20]. Obtaining an ECG should be considered, as it may identify early myocardial injury [20]. Therefore, removal of the patient from the environment and initiation of normo-baric oxygen therapy in the field should be considered followed by coordination with emergency departments and escalation to a facility with hyperbaric capabilities for those with moderate to severe signs of CO poisoning (Table 5.4).

Table 5.4 Flash-to-bang technique

IF thunder is heard	The lightning is
5 seconds after a flash	1 mile away
10 seconds after a flash	2 miles away
15 seconds after a flash	3 miles away
20 seconds after a flash	4 miles away
25 seconds after a flash	5 miles away
30 seconds after a flash	6 miles away
35 seconds after a flash	7 miles away
40 seconds after a flash	8 miles away

5.6 Extraction

Effective extraction and emergency response during stormy conditions require meticulous planning and preparedness, extending beyond medical personnel to include event administrators. There are two primary strategies for extraction: evacuating individuals from the field to nearby shelters and, if necessary, transporting them to hospitals. Preparedness is a crucial factor for the administration of sports events under such conditions.

In extreme weather conditions, it is advisable to prohibit spectator attendance to reduce the burden on authorities responsible for ensuring everyone's safety. In scenarios where preparation is inadequate, priority should be given to evacuating vulnerable groups, such as children, pregnant women, and the elderly, to nearby shelters first.

Each venue presents unique challenges for the safe extraction of athletes and spectators. The team physician must be aware of all potential obstacles that could impede the transport of victims and uninjured individuals to safety. Extracting victims during severe weather conditions is complex, particularly when CPR or automated external defibrillator (AED) use is required during transport. Injured spectators can be more challenging to manage and extract compared to athletes.

In the event of thunderstorms and lightning, a reverse triage system is employed. Activities must be suspended until 30 minutes after the last lightning strike is seen or heard. As part of the Emergency Action Plans (EAP), safe locations and buildings should be identified. It is imperative to move victims to a controlled setting. Fully enclosed buildings with electrical wiring and plumbing are the safe location of choice. However, fully enclosed vehicles such as ambulances, school buses, cars, or vans may be used instead if the event is taking place in an outdoor venue and enclosed buildings are not readily available. Spinal precautions should be maintained as even transient paralysis may mimic spinal cord injury. Therefore, these victims are considered trauma patients. The physician should be mindful that weather conditions may impede air transport, so ground transportation may be the transportation of choice [19, 21].

In the setting of tornado watch, sporting events should be canceled in the setting of tornado watch. If a tornado warning is issued prior to the start of the event, the event should also be canceled. If the tornado warning is issued during the event, participants should be led to prior designated tornado shelter. Rescue attempts during tornadoes place the first responder and rescuers at risks. In fact, 50% of tornado-related injuries during a tornado in Marion, Illinois, were registered during rescue attempts [13].

For all severe weather scenarios, first responders must provide immediate care while ensuring their own safety. Effective communication is vital for successful extractions. Regular drills and simulations can help familiarize all involved parties with evacuation procedures. In stormy conditions, two-way radios (walkie-talkies) offer reliable [22], instant communication essential for coordinating evacuations and emergency responses. Unlike cell phones, walkie-talkies provide immediate, real-time communication without relying on network signals, ensuring consistent connectivity even during power outages. This allows event organizers to efficiently manage the movement of athletes and spectators to safety and provide timely updates to all team members, enhancing overall safety and coordination. The new innovations in two-way radio communication system during the mass gathering are promising [22].

Special attention should be paid to handing off patients to medical evacuation staff. Clear and effective communication is crucial to ensure timely care and appropriate placement. Sedgley and colleagues proposed the mnemonic MIST for handoff, which is as follows [23]:

MIST mnemonic for handoff of patients by Sedgley et al. [23]	
M	Mechanism of injury
I	Injuries identified
S	Signs/symptoms
T	Treatment received

5.7 Prevention

Prevention of stormy weather-related hazards during sports events requires meticulous planning and preparedness to manage medical emergencies. Proper preparation involves creating comprehensive EAPs that include communication plans, designation of an incident commander, and coordination of additional resources [24]. These plans must consider various scenarios, such as natural disasters, to ensure prompt access to emergency personnel. The Incident Command System (ICS) is a vital organizational tool for managing responses to emergencies, involving command, control, and coordination among multiple agencies to effectively protect life and property [25].

Additionally, medical teams must be equipped with basic trauma life support skills, essential for the initial management of trauma patients. Triage is critical for sorting victims by need during mass casualty incidents caused by severe weather conditions [26]. The team physician's role is crucial in disaster scenarios, involving organizing the initial response, performing triage, and delivering basic trauma life support [27]. Besides these, various types of stormy weather-related preventive measures are described below. By implementing these measures, sports events can ensure a higher level of safety and preparedness, regardless of the specific nature of the emergency.

It is important to review weather forecasts prior to every game. If electric thunderstorms are expected, a strong consideration should be given to postponing sporting events [13].

The sports medicine clinician should understand the differences between a tornado watch and a tornado warning. A tornado watch implies that weather conditions are favorable for tornadoes, so careful monitoring and preparedness plans should be put in place to anticipate severe weather. On other hand, a tornado warning implies a tornado has been reported by weather spotters. It is imperative to find a safe shelter [5].

In the event of power outages caused by natural disasters such as hurricanes, we recommend having battery-powered CO detectors inside your place of shelter.

5.7.1 Lightning

Preventing lightning risk during sports activities, especially in outdoor venues like golf courses, requires thorough planning and real-time monitoring. Implementing a comprehensive lightning safety plan is essential, involving access to lightning proximity data through detection systems [28, 29] or commercial notification services, and continuous updates from the National Weather Service (NWS) [30]. A designated weather watcher should monitor conditions using various sources such as television, the internet, NOAA weather radio, and mobile alerts.

Sports activities should be halted when lightning is detected within specific radii: 15 miles for monitoring, 12 miles for caution, and 8 miles for immediate action [30]. For sporting venues, all activities should be halted until 30 minutes after seeing a lightning strike, hearing a thunder, or seeing thunderclouds [8, 18, 31]. There is no absolutely safe place from lightning—some locations are safer than others. When possible, shelter should be sought in the largest enclosed building available away from doors or windows. Another option is in a metal-topped vehicle with windows and doors closed; convertibles with fabric tops are not protective [32].

Effective communication is crucial; use public address systems, text/email alerts, social media, and staff announcements to inform staff of the lightning threat and direct them to safety. Providing education on lightning safety protocols to both staff and athletes ensures everyone knows how to respond to a threat. Additionally, employing a "reverse triage" strategy, where priority is given to those without vital signs or spontaneous respirations, is recommended during multiple lightning casualties to maximize survival rates [19]. By adhering to these guidelines and maintaining a proactive stance, the risk of lightning-related injuries during outdoor sports activities can be significantly mitigated.

Golf courses, with their open fields and isolated trees, are particularly vulnerable to lightning strikes. They must adhere to stringent safety protocols [33], including installing a reliable lightning detection system and subscribing to commercial lightning notification services. Golf carts and partially enclosed shelters are unsafe. Open shelters (such as dugout or canopy) should be avoided because of the risk of side splash and ground current [19]. When possible, the safest shelters are a building followed by a hardtop vehicle. Tents are not safe shelters [34].

The safety plan must include clear guidelines for evacuating the course. Substantial shelter buildings should be identified with clear signage indicating their locations [35, 36]. Communication with golfers is critical; use public address systems, notifications in golf carts, and printed safety guidelines on scorecards to inform and direct golfers to safety. By implementing these measures, golf courses can significantly reduce the risk of lightning-related incidents and ensure a safer environment for patrons.

Public notification should involve multiple channels such as outdoor warning sirens, public address systems, internal television/radio broadcasts, and social media alerts. Clear signage must direct to designated lightning-safe structures, and evacuation plans should be rehearsed regularly. Investing in lightning protection equipment that intercepts and diverts lightning away from populated areas is also essential. The Wilderness Medical Society recommends separating group members by at least 20 feet during a storm to prevent mass casualties, as lightning can jump between individuals [19].

The lightning position should be taught to everyone in the team. It involves crouching or sitting with knees and feet close together to create a single contact point with the ground (Fig. 5.2). If standing, keep feet touching; if sitting, lift feet off the ground. Use this position only when a lightning strike is imminent and other prevention strategies have failed. Signs of an imminent strike include St. Elmo's fire, static electricity, an ozone smell, or a nearby crackling sound [21, 22]. To minimize ground current injury, insulate yourself from the ground by sitting on a pack (without metal), a dry coiled rope, or a rolled foam pad. This is a last-resort strategy, not primary prevention, and is hard to maintain for long periods. By ensuring these measures, outdoor venues can provide a safer environment during lightning events [37, 38].

Fig. 5.2 Lightning strikes at the US High Performance Center of Tokyo Olympic Games 2021

5.7.2 Heavy Rain and Flooding

Heavy rain and flooding are often accompanied by lightning [39], necessitating the implementation of comprehensive lightning safety measures. These include real-time weather monitoring, alerting individuals in the stadium, and providing nearby shelters. Additionally, having backup power supplies and pre-event evacuation plans in place is crucial.

The most effective preventive measure is to either cancel or postpone the sports event during such adverse weather conditions. However, in case of sudden weather changes, preparedness for cold-related injuries is also important. Cold weather can slow down the body's chemical processes, including the nervous system's ability to generate muscle contractions. Activities on damp grounds pose significant injury risks due to slipping hazards, especially in sports such as baseball, cricket, golf, and tennis [40]. Poor visibility under these conditions can also make sports dangerous or disadvantageous for competitors.

Medical teams should be prepared to treat hypothermia-related injuries among athletes and on-ground support staff during heavy rain and flooding [41]. Prolonged exposure to heavy rain can lower core body temperatures, necessitating immediate intervention by providing dry clothing, warm liquids, or warm showers to treat hypothermia. Nonfreezing tissue injuries of the extremities, such as trench foot [41], should be managed expediently, as these are caused by prolonged exposure to cold, wet conditions, unlike frostbite which results from excessively cold but dry environments.

Athletes should wear appropriate clothing and gear, such as waterproof and windproof layers, to maintain body temperature and comfort during outdoor activities. Additionally, it is crucial for athletes to stay hydrated in cold weather, even if not thirsty, to avoid dehydration caused by increased urine production and evaporation [41].

Artificial grass pitches, like AstroTurf, can handle more usage in wet conditions compared to natural grass but still have their drawbacks, such as causing the ball to skid when wet [40].

Implementing these preventive measures will help ensure the safety and performance of athletes during heavy rain and flooding, thereby minimizing the risk of injuries and related complications.

5.7.3 Tornado and Hurricanes

Typhoons, hurricanes, and tornadoes are all categorized as tropical storms. A hurricane is defined as a tropical system with sustained winds reaching a minimum speed of 74 miles per hour [42]. The eye of a hurricane typically spans 20–30 miles in diameter, while the overall system can extend over 400 miles in width. The primary hazards associated with these storms include torrential rains, high winds, storm surges, and tornadoes, all of which involve violent wind conditions.

To prevent injuries related to high winds during sports events, it is crucial to prioritize structural safety and the securement of equipment and gear. All temporary structures, such as tents, stages, and seating areas, must be securely anchored using heavy-duty stakes, sandbags, or concrete blocks to ensure stability during high winds. Regular inspections and necessary reinforcements of these structures are essential. Additionally, any loose objects or equipment that could become hazardous projectiles must be securely anchored or removed to prevent injuries [36].

Athletes should be equipped with appropriate protective gear that does not pose a hazard in high winds. Continuous monitoring of weather conditions is imperative, along with clear communication of emergency plans, identification of safe zones, and readiness to postpone or cancel events if necessary. Medical teams should be prepared for immediate response to any injuries, ensuring the safety of all participants, staff, and spectators. By adhering to these preventive measures, the risks associated with heavy winds during tropical storms can be significantly mitigated (Table 5.5).

5.7.4 Tornadoes

In the event of a tornado, move victims to predesignated safe shelters such as basements or interior rooms on the lowest floor, away from windows. Avoid vehicles and open structures. If no shelter is available, lying flat in a nearby ditch or depression and covering the head with the arms is recommended.

During extraction, Be ready to provide emergency treatment, including CPR and maintaining spinal precautions.

Table 5.5 Quick reference: Lightning protocols/policies by the organization

Lightning protocols/policies by the organization	
Organization	Policy
National Athletic Trainers' association	Implementation of a lightning-safety policy that identifies safe locations for shelter from the lightning hazard Monitoring local weather forecasts, designating a weather watcher, and establishing a chain of command Flash-to-bang count of 30 seconds or more should be used as a minimal determinant of when to suspend activities Waiting 30 minutes or longer after the last flash of lightning or sound of thunder is recommended before athletic or recreational activities are resumed [7]

(continued)

Table 5.5 (continued)

Lightning protocols/policies by the organization	
Organization	Policy
National Collegiate Athletics Association	Education and prevention are the keys to lightning safety Prevention: Develop a lightning safety plan for each outdoor venue 1. Designate a person to monitor threatening weather and to notify the chain of command 2. Monitor local weather reports daily 3. Monitor local weather reports daily 4. Be informed of National Weather Service (NWS)-issued thunderstorm watches or warnings and know the difference between a severe thunderstorm watch and a severe thunderstorm warning 5. Know where the closest "safer structure or location" is to every outdoor venue and know how long it will take to evacuate everyone to that safer location 6. Lightning awareness should be heightened at the first flash of lightning, clap of thunder, and/or other signs of an impending storm, such as increasing winds or darkening skies, no matter how far away. These types of activities should be treated as a warning or wake-up call to athletics personnel 7. Use the flash-to-bang method to estimate the distance between your location and the lightning flash 8. To resume athletics activities, lightning safety experts recommend waiting 30 minutes after both the last sound of thunder and after the last flash of lightning is at least eight miles away and moving away from the venue 9. Play should be suspended if there is lightning within 8 miles of the stadium [1, 8]
PGA tour	Officials are constantly monitoring the weather with an on-site meteorologist and state of the art technology In the event that lightning is detected at a distance from the course that threatens the safety of the individuals at the event, play is immediately suspended, and the course is evacuated for safety precautions The distance from the course however is not explicitly documented [24]
US rugby	Obtain a weather forecast before an outdoor event in addition to knowing your local weather patterns. Identify and stay within traveling range of a proper shelter Employ the "30–30 rule" to know when to seek a safer location [25]
US soccer federation (USSF)	If you can hear thunder, you are within reach of lightning. Referees must protect the safety of all participants by stopping game activities quickly, so that participants and spectators may retire to a safer place before the lightning threat becomes significant Apply the 30–30 rule and wait 30 minutes or more after hearing the last thunder before leaving shelter [27]
US equestrian and the interscholastic equestrian association	Having a plan and a safe place to shelter in the event of lightning Outdoor events be suspended or moved indoors if a storm is less than six miles away (including mounting and warm-ups) [28] In order to determine the distance of a storm in miles, use the flash-to-bang method by counting the number of seconds between seeing lightning and hearing thunder and dividing it by five (lightingsafety.noaa.gov) [20]
MLB	Home team shall be the sole judge as to whether a game shall not be started because of unsuitable weather conditions or the unfit condition of the playing field, except for the second game of a conventional or split doubleheader [29] No specific lightning guidance or protocols are outlined as it is up to the home team to determine safety and whether game should be started

(continued)

Table 5.5 (continued)

Lightning protocols/policies by the organization	
Organization	Policy
US open water swimming	"2019 plan: The following priority will occur in this order, if the safety officer, meet referee, meet director, or independent safety monitor becomes aware of any situation requiring abandonment (including inclement weather) 1. Radio communication to officials and safety personnel on course 2. All lifeguards issue a three loud short blast whistle command and instruct all swimmers to evacuate the course 3. Announcer instructs all swimmers, safety personnel, and officials, to exit the course. If the reason for abandonment is inclement weather, announcer will instruct all athletes, coaches, parents, and meet personnel to go to the shelter, team bus, or vehicles, until inclement weather passes 4. Everyone will be instructed to go to vehicles, until safe conditions return" [30] No specific distance regulations or detection methods outlined
US tennis	Lightning is a potential severe hazard and life-threatening consequence of an approaching storm near outdoor tennis matches. Important to be prepared for immediate cessation of all matches or warm-up in the event of lightning If lightning is sighted, stop all activity and direct everyone to seek appropriate shelter A 30–30 rule may be used [31]
US beach volleyball	"The weather must not present any danger of injury to the players" [32] No specific guidelines noted
The USOPC (US Olympic and Paralympic committee)	No longer has information regarding lightning safety or severe weather protocols listed on its website. No policy or protocol found [33]
IOC (International Olympic Committee)	No longer has information regarding lightning safety or severe weather protocols listed on its website. No policy or protocol found [34]
World Rugby	Live data: In circumstances, where live data is available from local meteorological services along with advice from an expert, this data should be used to assist in the decision about whether to allow the event/match to commence, continue, or be abandoned Determination of a chain of command which clearly identifies the person with ultimate responsibility to stop the event, either temporarily or indefinitely, or to evacuate the venue is vital. It should be ensured that this person is familiar with local protocols and is provided with accurate and up-to-date information Communication with local weather services or tracking of weather alerts should be established in advance of any event When the lightning strike is within 6 miles of the venue, it is recommended that players are removed from the pitch [35] Flash-to-bang technique to determine distance when live data direct from local weather services is available Recommended safe distances from the lightning event is 6 miles For events/venues where live data direct from local weather services is not available, the 30–30 rule should be used to ensure participant and spectator safety: • *Shelter should be sought when there are 30 seconds or less between the flash and the associated thunderclap.* • *Safe to return to the field of play is considered only after a minimum of 30 minutes after the final flash of lightning or clap of thunder has been seen/heard.* The 6-mile recommendation may be altered depending on the topography of the area, for example, to increase this distance in flat open areas such as deserts or decrease it if the storm is located on the other side of a mountain range For spectators, individual stadium safety protocols should be followed in the event of a strike risk. For events which are not taking place in a stadium, the same protocols as those for players should be followed [35]
FIFA	No longer has information regarding lightning safety or severe weather protocols listed on its website. No policy or protocol found [36]

5.8 Take-Home Pearls

1. Conduct a thorough pre-event weather assessment in anticipation of any event. Remember to have a designated person continuously monitoring the weather during the event.
2. Review security and disaster protocols. Be familiar with the emergency action plan.
3. In the event of a disaster, establish a clear chain of command.
4. Be familiar with shelter locations in the vicinity of your event.
5. For lightning victims, a reverse triage system is employed. Care is prioritized to unconscious victims, as those conscious are likely to survive.
6. It is safe to touch a lightning strike victim.
7. Sudden death from lightning strikes involves both cardiac and respiratory arrests. It is necessary to perform high-quality CPR while securing and maintaining ventilation.
8. Responders should be ready to re-resuscitate lightning victims.
9. Responders should be aware of possible injuries during rescue and extraction. Well-trained personnel should be in charge of rescue and extraction.

References

1. CDC. Lightning. 2024 [cited 2024 May 28]. Lightning strike victim data. Available from: https://www.cdc.gov/lightning/data-research/index.html.
2. US Department of Commerce NO and AA. What is a hurricane? [Internet]. [cited 2024 Jun 2]. Available from: https://oceanservice.noaa.gov/facts/hurricane.html.
3. Saffir-Simpson Hurricane Wind Scale [Internet]. [cited 2024 June 2]. Available from: https://www.nhc.noaa.gov/aboutsshws.php.
4. Blake ES, Landsea CW, Gibney EJ. NOAA Technical Memorandum NWS NHC-6.
5. NOAA National Severe Storms Laboratory [Internet]. [cited 2024 June 9]. Tornado Basics. Available from: https://www.nssl.noaa.gov/education/svrwx101/tornadoes/.
6. 25 Deadliest U.S. Tornadoes [Internet]. [cited 2024 June 9]. Available from: https://www.spc.noaa.gov/faq/tornado/killers.html.
7. Thomson EM, Howard TM. Lightning injuries in sports and recreation. Curr Sports Med Rep. 2013;12(2):120.
8. Bauer AK, Golden KG, Colvin CM, Lammlein KP, Wise SR. When lightning strikes: sports and recreational activities safety. Curr Sports Med Rep. 2023;22(4):126.
9. Ritenour AE, Morton MJ, McManus JG, Barillo DJ, Cancio LC. Lightning injury: a review. Burns. 2008;34(5):585–94.
10. Norman ME, Albertson D, Younge BR. Ophthalmic manifestations of lightning strike. Surv Ophthalmol. 2001;46(1):19–24.
11. O'Keefe Gatewood M, Zane RD. Lightning injuries. Emerg Med Clin North Am. 2004;22(2):369–403.
12. Hurricane | Impact [Internet]. [cited 2024 July 17]. Available from: https://community.fema.gov/ProtectiveActions/s/article/Hurricane-Impact.
13. Woodward T, Shamim F, Hinson M, Bass E. Unexpected disasters at organized sporting events: considerations in preparation and response. Curr Sports Med Rep. 2015;14(3):171.
14. Centers for Disease Control and Prevention, USA. Carbon monoxide poisoning after hurricane Katrina--Alabama, Louisiana, and Mississippi, August–September 2005. MMWR Morb Mortal Wkly Rep. 2005;54(39):996–998.
15. Iqbal S, Clower JH, Hernandez SA, Damon SA, Yip FY. A review of disaster-related carbon monoxide poisoning: surveillance, epidemiology, and opportunities for prevention. Am J Public Health. 2012;102(10):1957–63.
16. Eichner ER. Accidental carbon monoxide poisoning in sports and other settings. Curr Sports Med Rep. 2018;17(3):78.
17. Rose JJ, Wang L, Xu Q, McTiernan CF, Shiva S, Tejero J, et al. Carbon monoxide poisoning: pathogenesis, management, and future directions of therapy. Am J Respir Crit Care Med. 2017;195(5):596–606.
18. Walsh KM, Cooper MA, Holle R, Rakov VA, Roeder WP, Ryan M. National Athletic Trainers' association position statement: lightning safety for athletics and recreation. J Athl Train. 2013;48(2):258–70.
19. Davis C, Engeln A, Johnson EL, McIntosh SE, Zafren K, Islas AA, et al. Wilderness medical society practice guidelines for the prevention and treatment of lightning injuries: 2014 update. Wilderness Environ Med. 2014;25(4_suppl):S86–95.
20. Carbon Monoxide Poisoning [Internet]. [cited 2024 June 9]. Available from: https://www.acep.org/patient-care/clinical-policies/carbon-monoxide-poisoning.
21. A 26-Year-Old Man Struck by Lightning – ClinicalKey [Internet]. [cited 2024 July 17]. Available from: https://www-clinicalkey-com.proxy.cc.uic.edu/#!/content/playContent/1-s2.0-S1067991X14003113?scrollTo=%23hl0000185.

22. Lien YN, Chi LC, Huang CC. A multi-hop walkie-talkie-like emergency communication system for catastrophic natural disasters. In: 2010 39th International Conference on Parallel Processing Workshops [Internet]. San Diego, CA, USA: IEEE; 2010 [cited 2024 July 22]. p. 527–32. Available from: http://ieeexplore.ieee.org/document/5599115/.
23. Sedgley M, Hudson K, Hulsopple C. Prepare for the unexpected: a new look at trauma triage and care in mass participation sporting events. Curr Sports Med Rep. 2023;22(1):4.
24. Aaron R. Sports injuries and emergencies: a quick response manual. New York: McGraw-Hill, Health Professions Division; 2003. (Sports injuries & emergencies).
25. Rubin AL. Safety, Security, and preparing for disaster at sporting events. Curr Sports Med Rep [Internet]. 2004;3(3). Available from: https://journals.lww.com/acsm-csmr/fulltext/2004/06000/safety,_security,_and_preparing_for_disaster_at.6.aspx.
26. Streger MR. Prehospital triage. Emerg Med Serv. 1998;27(6):21, 23–7, 45 passim.
27. Campbell JE, American College of Emergency Physicians, editors. Basic trauma life support for paramedics and other advanced providers. 4th ed. Upper Saddle River: Brady/Prentice Hall Health; 2000. 364 p.
28. An Overview of Lightning Detection Equipment – National Lightning Safety Institute [Internet]. [cited 2024 June 16]. Available from: http://diliev.com/Home/applications/Library/Lightnings/lightning-safety5/www.lightningsafety.com/nlsi_lhm/detectors.html.
29. Overview of real-time lightning detection systems and their meteorological uses [Internet]. [cited 2024 June 16]. Available from: https://repository.library.noaa.gov/view/noaa/11301.
30. US Department of Commerce N. Lightning Safety and Outdoor Sports Activities [Internet]. NOAA's National Weather Service; [cited 2024 June 16]. Available from: https://www.weather.gov/safety/lightning-sports.
31. CDC. Lightning. 2024 [cited 2024 July 15]. Safety guidelines. Available from: https://www.cdc.gov/lightning/safety/index.html.
32. 5.4 Lightning-caused deaths and injuries in the vicinity of vehicles (2008 – 88Annual_3lightning) [Internet]. [cited 2024 June 16]. Available from: https://ams.confex.com/ams/88Annual/techprogram/paper_135740.htm.
33. US Department of Commerce N. Toolkits for County, Communities, Organizations and Large Venues [Internet]. NOAA's National Weather Service; [cited 2024 June 16]. Available from: https://www.weather.gov/safety/lightning-toolkits.
34. Zack F, Hammer U, Klett I, Wegener R. Myocardial injury due to lightning. Int J Legal Med. 1997;110(6):326–8.
35. Zimmermann C, Cooper MA, Holle RL. Lightning safety guidelines. Ann Emerg Med. 2002;39(6):A1.
36. Cooper MA, Holle R, Lopez R. Recommendations for lightning safety. JAMA. 1999;282(12):1132–3.
37. Lightning Safety | Maine Emergency Management Agency [Internet]. [cited 2024 June 16]. Available from: https://www.maine.gov/mema/hazards/natural-hazards/lightning-safety.
38. JM M. St. Elmo's Fire. Nature. 1912;89(2210):7–8.
39. Mudiar D, Pawar SD, Hazra A, Gopalkrishnan V, Lal DM, Chakravarty K, et al. Lightning and precipitation: the possible electrical modification of observed raindrop size distributions. Atmospheric Res. 2021;259:105663.
40. Thornes J. The effect of weather on sport. Weather. 1977;32:258–68.
41. Cappaert TA, Stone JA, Castellani JW, Krause BA, Smith D, Stephens BA, et al. National Athletic Trainers' Association position statement: environmental cold injuries. J Athl Train. 2008;43(6):640–58.
42. US Department of Commerce NO and AA. What is the difference between a hurricane and a typhoon? [Internet]. [cited 2024 July 17]. Available from: https://oceanservice.noaa.gov/facts/cyclone.html.

It's Rugged and Isolated

6

Victoria E. Bergstein, Eve R. Glenn, Brienna K. Buchanan, Henry T. Shu, Jacob D. Mikula, Matthew J. Best, and Stephen Weber

6.1 Introduction

High-altitude mountainous regions, defined as land areas situated over 600 meters (m) above surrounding terrain, boast steep topography and rapid changes in altitude over short horizontal distance. Altitude categories range from low (<1500 m) to moderate (1500–2500 m), high (2500–3500 m), or very high (3500–5000 m), with sea level as the reference point [1–3]. The burgeoning global interest in high-altitude sports has become increasingly evident, as the 2022 Olympic Winter Games garnered over 2.01 billion viewers [4, 5]. However, the growing popularity of mountain sports may be accompanied by an increased risk of injuries [6, 7].

Mountainous environments pose unique challenges to athletes, including steep slopes, unpredictable weather conditions, reduced atmospheric oxygen pressure, and technical navigation complexities. These challenges not only contribute to an elevated risk of accidents and injuries but may also increase difficulty of access for rescue services, which can delay treatment. Understanding these hazards is crucial for athletes, coaches, event organizers, and healthcare providers to ensure the well-being of participants. This chapter explores the unique risks of mountain sports and prioritizes the analysis of injury profiles, prevention strategies, and injury management within this challenging terrain.

6.2 Case Anecdote

Summiting Mount Everest, the highest point on Earth at an elevation of 8848 m (29,028 feet), is a dangerous task, once reserved only for the world's top mountaineers. However, in 1993, the allure of standing on top of the world helped launch a commercial industry where seasoned tour guides and Nepali Sherpas would lead inexperienced climbers up the famed peak (Fig. 6.1) [8].

In May 1996, a journalist named Jon Krakauer embarked on an expedition to summit Mount Everest [9]. Krakauer began his journey with a group of professional guide climbers, two Sherpas, and other amateur climbers like himself. As he recounts in his memoir, *Into Thin Air*, five of the original eleven members of his group never made it back down the mountain, which he terms the "Everest disaster."

Krakauer's account of his Everest expedition is an extreme and vivid depiction of the incremental effects of altitude sickness. At Base Camp, 5364 m above sea level, Krakauer experienced a migraine so severe he thought he was

V. E. Bergstein (✉) · E. R. Glenn · B. K. Buchanan · H. T. Shu · J. D. Mikula · M. J. Best · S. Weber
Johns Hopkins University Department of Orthopaedic Surgery, Baltimore, MD, USA
e-mail: vbergst2@jhmi.edu; eglenn8@jhmi.edu; bbucha10@jhmi.edu; hshu5@jhmi.edu; jmikula2@jhmi.edu

M. R. Hutchinson, S. Rocha Piedade (eds.), *Sports Medicine: It's Environmental*,
https://doi.org/10.1007/978-3-031-84493-5_6

Fig. 6.1 Mountaineers climbing the Hillary Step, a challenging portion of the ascent, near the peak of Mount Everest. (Photo Credit: Laxman Karki, Mountain Ascend Treks)

having a stroke. This effect was likely due to acute high-intensity ultraviolet (UV) radiation exposure and subsequent UV-induced photokeratitis. Krakauer attested that he felt like the UV radiation had "burned [his] retinas and baked [his] brain", forcing him to lie in his tent "trying to avoid sensory stimuli of any kind" [9 p58].

One of the Sherpas in Krakauer's group, Ngawang, developed high-altitude pulmonary edema (HAPE) at Camp II at approximately 6500 m. HAPE is a rare condition known to affect healthy climbers within 2 to 5 days of arrival at altitudes above 2500 to 3000 m [10]. The only definitive cure for HAPE is a rapid descent from altitude of at least 300–1000 m or until the patient is asymptomatic [11]. In attempts to avoid descent, Ngawang was administered acetazolamide, dexamethasone, and sublingual nifedipine, given supplemental oxygen, and placed into an inflatable hyperbaric chamber, but none of those interventions proved effective. Following his reluctant but necessary descent, Ngawang died from complications of HAPE within 2 months of onset [9 p74–75].

In contrast to HAPE, high-altitude cerebral edema (HACE) is less common but tends to be more deadly. A climber on Krakauer's expedition named Dale Kruse was struck by HACE at Camp III, at an altitude of over 7100 m. Kruse described the experience, reporting that he laid down for a nap "and ended up sleeping almost twenty-four hours… It was like I was very drunk… I couldn't walk without stumbling, and completely lost the ability to think or speak" [9 p88–89]. Luckily, Kruse was immediately evacuated to base camp, where his symptoms resolved within 3 or 4 days.

As a storm raged on Everest's peak during the climbers' descent, extreme elements battered the slower climbers. Beck Weathers spent 12 hours in a comatose state [9 p145], buried in snow and left for dead by others on the expedition before regaining consciousness and finding his way back to Camp IV. Despite narrowly escaping an icy death, medical examination revealed a horrifying degree of frostbite necessitating several amputations. Krakauer reports that "all four fingers and the thumb on [Weather's] left hand were removed. [Additionally,] his nose was amputated and reconstructed with tissue from his ear and forehead" [9 p160].

The infamous "Everest disaster" should be taken as a cautionary tale. Nevertheless, as Krakauer notes in his novel, the thrill sought by mountaineers and adventurers continues despite the risks; even after learning of the tragedy that occurred in Krakauer's group, the remaining expeditions on the mountain continued with their ascent [9 p157–159].

While the medical conditions described by Krakauer represent some of the most severe examples of sports-related injuries at extreme altitude, these conditions may also present at relatively lower altitudes. The remainder of the chapter will discuss the epidemiology, unique injuries, management, extraction challenges, and prevention of sports-related injuries at altitude.

6.3 Epidemiology

For the over 35 million people that visit high-altitude destinations over 3000 m every year, exposure to high altitude warrants special health considerations [12]. Understanding the unique risks of various high-altitude sports is crucial for these athletes and the medical teams that treat them.

6.3.1 Climbing

Climbing is a multifaceted sport that encompasses alpine climbing, sport or traditional climbing, and ice climbing. With an injury rate ranging from 0.56 to 37.5 injuries per 1000 hours of climbing, one thing is clear: injury risk in climbing is highly variable, and individual climbers, environments, and climbing subtype must all be considered [13].

Alpine climbing combines the sports of rock climbing and mountaineering, and involves ascending rock faces at high altitude. Alpine climbing proves to be especially perilous in the winter, with the majority of accidents resulting from falls on rock, or on snow and ice [14]. In terms of fall injuries, the three most common are lower extremity fractures, rib fractures, and craniofacial fractures [14]. Following fractures, abrasions/contusions and ligamentous injuries are second and third most common, respectively [15]. Head injuries account for approximately 20% of all injuries, fall-related and otherwise, highlighting the importance of safety equipment such as helmets. Climbers are also at risk for injury due to falling debris, representing 30% of climbing accidents in a study conducted at Grand Teton National Park [14]. Alpine climbers should be especially wary of rock and snow hazards, and wearing a helmet can prevent serious injury.

Sport climbing and traditional climbing involve ascending rock faces with minimal equipment toward fixed anchors to clip in for protection against falling. While alpine climbing injuries are most often attributable to falls, sport climbing injuries are primarily associated with executing strenuous moves [13]. Despite the majority of climbing injuries being classified as of minor severity, climbing still has a risk of fatality, particularly in snowy or icy environments where falls were found to be further and injuries more likely to be fatal [16].

Ice climbing involves ascending frozen waterfalls, ice slabs, and frozen-over rocks using ice axes and crampons. Although there are minimal published data describing the incidence of ice climbing injuries, one prospective study following 70 ice climbers found that ice climbing had an injury rate of 9.8 injuries per 1000 hours of climbing [17]. Intermediate-level ice climbers have 2.6 times greater odds of injury risk compared with advanced ice climbers, indicating that climber ability plays a major role in injury risk and should potentially be considered in prevention efforts [17]. Regions of the body most often injured while ice climbing include the head, making up nearly half of all injuries, followed by the knee and shoulder [17]. Abrasions and contusions are the most widely reported injury in ice climbing, and falling ice is the leading cause of injury, which makes wearing helmets during this activity of utmost importance [17].

While the environment is a key factor in injury risk during climbing, athlete decision-making has also been demonstrated to be a contributor to injury risk. An estimated 92% of rock climbing falls involved the lead climber, often due to attempts at skills beyond their abilities, loose rock, longer falls compared to climbers behind the lead, and potential for protection point failure [15]. Despite this predominance, fall risk has a bimodal distribution with spikes occurring among both expert climbers and inexperienced climbers [14]. It has been postulated that novice climbers are at increased risk due to lack of skill, while experienced climbers are more likely to fall as a result of testing the limits of their abilities.

6.3.2 Mountaineering

Mountaineering is a combination sport that challenges athletes to traverse terrain and summit mountains of varying elevations using multiple modalities ranging from climbing to trekking. Injuries occur at a rate of approximately 5.6 injuries per 10,000 hours of mountaineering, although risk is highly variable due to the wide range of mountains encapsulated within the mountaineering sport [13, 18]. Lack of proper equipment was associated with 14.3 times higher odds of injury in mountaineering [18].

While many injuries associated with high altitude often involve fractures and ligamentous injuries, musculoskeletal injuries appear to be less common among mountaineers. In a study of mountaineering injuries at Denali National Park, altitude syndromes made up the vast majority of complaints, followed by frostbite and hypothermia [19].

6.3.3 Mountain Biking

Mountain biking has risen in popularity in recent years and attracted millions of athletes to the sport [20]. Subtypes of mountain biking include cross-country and trail riding, downhill biking, and freeride. The subtype of mountain biking is important in risk assessment, with downhill biking representing 85.6% of severe acute injuries, often via a falling-forward mechanism [20, 21]. Injuries in cross-country mountain biking, conversely, often involve collisions with other cyclists [22]. Rates of injury also vary with subtype and sex, ranging from 3.1 injuries per 1000 hours for male cross-country cyclists to 46.8 injuries per 1000 hours for female downhill cyclists [20].

Despite the differences between subtypes, there are several common injury patterns across all mountain biking disciplines. Approximately two-thirds of acute injuries occur in the upper extremity, while the same fraction of overuse injuries occur in the lower extremity [20]. Female mountain bikers have higher rates of injury compared to their male counterparts, although male mountain bikers make up the majority of athletes and represent 86% of mountain biking injury presentations to the emergency room [20]. Soft tissue injuries are most common overall, representing 60–75% of injuries [23]. Up to 75% of fractures occur in the upper extremities, especially the clavicle, and acromioclavicular joint separations are also common among all types of mountain bikers [20, 23].

One aspect of injury prevention unique to mountain biking is bike maintenance. Bike maintenance and fitting are essential to limiting the proportion of injuries attributable to equipment failure, currently estimated to make up between 6% and 16% of injuries [22].

6.3.4 Skiing and Snowboarding

Skiing and snowboarding represent some of the most popular high-altitude activities, with an estimated 200 million skiers and 70 million snowboarders on the slopes annually [24]. With a relatively low injury risk of 1–4 injuries per 1000 participant days, recreational snow sports are perhaps surprisingly safer than several sports at lower altitudes, including football and rugby [24].

Injury rates for skiing have not always been this low. Notably, skiing-related injuries have roughly halved since the 1940s, which may be attributable to improved safety rules, slope maintenance, and advancement of ski equipment [25]. In particular, the development of rigid ski boots has reduced ankle injury rates, and the addition of release bindings that enable skis to pop off when twisted has been instrumental in minimizing the force exerted on the involved lower extremity [25, 26]. Snowboarders have an estimated twofold higher risk of injury compared to skiers, especially in the upper extremity [24]. Interactions between skiers and snowboarders within a single resort environment can impact injury rates, with a 13% increase in injury rates seen in Taos Ski Valley upon the introduction of snowboarding alongside skiing [27].

6.4 Unique Illnesses and Injury Patterns

6.4.1 High-Altitude Illness

High-altitude illness (HAI) is the umbrella term for three syndromes acquired at high altitude, including acute mountain sickness (AMS), high-altitude cerebral edema (HACE), and high-altitude pulmonary edema (HAPE). HAI may occur beginning at elevations of 1500–2500 m, with the primary risk factor being a rapid rate of ascent without acclimatization [28]. Other risk factors include anemia, preexisting coronary artery or pulmonary disease, and permanent residence at low altitudes [28]. While physical fitness does not appear to contribute to or prevent HAI, exertion is thought to exacerbate HAI, meaning athletes training at high altitudes may be at an increased risk of having more severe symptoms [28].

AMS is the most common HAI and can occur within hours of ascent. At altitudes above 3800 m, AMS can occur in 50–84% of athletes [28]. The hallmark of AMS is headache and can be accompanied by other symptoms such as gastrointestinal upset, sleep disturbances, fatigue and weakness, and dizziness and lightheadedness. AMS is not associated with a high fatality rate, in contrast to the other forms of HAI [29]. The primary treatment for AMS is descent from altitude, but acetazolamide may also be used for treatment of mild disease or for patients who are unable to descend.

HAPE is a form of noncardiogenic pulmonary edema secondary to hypoxia at high altitudes [30]. In contrast to AMS, HAPE typically presents within 2–5 days of altitude exposure. Diagnosis of HAPE is primarily clinical and based on symptoms of respiratory distress, such as dyspnea on exertion progressing to dyspnea at rest, orthopnea, and cough productive of frothy pink sputum [30]. A standard chest radiograph showing peripheral permeability or overperfusion edema and normal pulmonary venous pressure, or the finding of bilateral pulmonary rales on auscultation, may be used to confirm the diagnosis but is often not available while the patient is at high altitude [31]. The key method of preventing HAPE is acclimatization or slow ascension over weeks to allow the body to gradually compensate for systemic hypoxia. Treatment of HAPE involves supplemental oxygen and immediate descent or placement in a hyperbaric chamber if descent is not possible. Oral nifedipine may be a useful adjunct in the treatment of HAPE [30, 32].

HACE is the rarest form of HAI and is characterized by ataxia, altered mental status, and fatigue. Like HAPE, it typically presents after 2 days of rapid ascent. The pathophysiology of HACE is hypothesized to be an exaggerated neurohormonal response to hypoxia, resulting in excessive cerebral vasodilation, edema, and intracranial hypertension. The "tight fit" hypothesis has postulated that variation in cerebrospinal fluid volumes and cerebral space are potential anatomic risk factors for HACE [12]. Like other HAIs, HACE is primarily a clinical diagnosis. On bloodwork, HACE has been associated with an elevated white blood cell count and elevated lumbar puncture opening pressure with normal cerebrospinal fluid. Head imaging, such as with a computed tomography scan and/or magnetic resonance imaging, may demonstrate cerebral edema. The cornerstone of HACE management is immediate descent of at least 1000 m or until symptoms improve. If descension is not available, supplemental oxygen and a hyperbaric chamber may be used as temporizing measures. Some studies have suggested that dexamethasone and acetazolamide may be useful in treating HACE [11, 33].

6.4.2 Other High-Altitude Environmental-Related Illnesses

Environmental factors also significantly contribute to injury patterns experienced in mountainous environments. Ultraviolet (UV) radiation exposure increases with altitude due to reduced atmospheric shielding, with a rise of 10–20% in UVA and UVB waves for every 1000 m of altitude [34, 35]. Photoprotective measures are essential at high altitudes; the reflective properties of snow and ice, in combination with the increased UV radiation exposure, predispose mountain athletes to more severe sunburn and photokeratitis than athletes at lower altitudes [36].

Extreme temperatures pose risks of injury, including heat stress, frostbite, and dehydration.

While external temperatures may be lower at high altitude, athletes may still experience heat stress secondary to wearing excessive thermal clothing during activity. Thus, high-altitude activities necessitate clothing choice that balances the risk of hypothermia from inadequate clothing while minimizing the risk of heat stress from excessive gear.

6.4.3 Snowboarding

Snowboarders are more prone to upper extremity injuries than skiers; because snowboards bind the feet together, fall on outstretched hand (FOOSH) injuries from axial loading are frequent mechanisms of injury [37, 38].

Another common mechanism of injury for snowboarders is falling backward onto the buttocks [39]. These backward falls, which can be caused by imbalance, collisions, or jumps, can result in fractures of the pelvic ring or isolated fractures of the pubic ramus, sacrum, and iliac wing [40, 41]. Fractures of the lateral process of the talus are also common in snowboarders, caused by axial loading followed by external rotation of the foot during aerial jumps or turns [42].

6.4.4 Alpine Skiing

There is a heightened occurrence of lower body injuries in alpine skiing. Isolated and multiligamentous knee injuries, especially of the ACL and MCL, are common, with one study finding ACL sprains to be the most common injury in skiers [43]. The most common mechanism for ACL tears in skiers is called "slip catch" [25, 44]. Slip catch occurs during a skier's turn, where the skier loses pressure on the outer ski, and in an attempt to regain traction on the snow, the ski gets caught, forcing the knee into internal rotation and valgus [45, 46]. Distal tibia and ankle fractures also occur due to valgus rotation and axial compression traumas. Individuals with novice and intermediate ski experience sustained more shaft area fractures, while those with higher experience levels showed a more balanced distribution of knee region injuries [47].

Falls in skiing are also a common culprit for shoulder injuries; the clavicle is involved in 32% of all skiing fractures [48]. In the hand, tears of the metacarpophalangeal ulnar collateral ligament—dubbed "skier's thumb"—can occur when one falls onto an outstretched thumb while grasping an object, such as a ski pole [49]. Skier's thumb has been estimated to represent between 1.5% and 2.3% of skiing injuries [50].

6.4.5 Cross-Country Skiing

In contrast to conventional downhill skiing, cross-country skiing bindings feature a free heel, connected to the ski solely at the toe. In cross-country skiing, hip hyperextension and femoral nerve injury can occur in the setting of one leg forcefully extending backward while the other slides forward to generate propulsion [43, 51, 52]. Cross-country skiing has similar injury profiles as alpine skiing and snowboarding, despite the notably lower sliding speed minimizing the severity of injury. In the upper extremities, compressive injuries including cubital tunnel syndrome, also known as ulnar neuropathy at the elbow, have been associated with forceful, repetitive poling [53].

6.4.6 Freestyle Skiing, Slopestyle Skiing, Ski Jumping, and Nordic Combined Skiing

Freestyle skiing involves navigating a course of moguls and jumps. Slopestyle skiing involves executing acrobatic tricks and maneuvers in terrain parks or halfpipes, while ski jumping challenges athletes to achieve the longest possible distance after launching off a ramp. Nordic combined involves ski jumping and cross-country skiing events. Common injuries include ligamentous knee injuries, head injuries, and lumbosacral spine injuries [54–56]. In freestyle skiing and ski jumping, high-velocity crashes from heights are a unique mechanism of injury [57].

6.4.7 Snowmobiling

Snowmobilers can reach speeds of up to 110 mph while operating 600-pound vehicles, thus making some accidents fatal [58]. Injuries can occur from collisions with fixed or moving objects, submersion into icy lakes, driving off cliffs, loss of control of the snowmobile, ejection, and rollovers [58, 59]. Snowmobiling injuries are most commonly seen in young males and are especially associated with alcohol consumption and riding at high speeds [58, 59]. Due to the dangerous nature of snowmobiling, polytrauma occurs at a higher rate than most other snow sports [60].

Head injuries in snowmobiling range from skull fractures to neck injuries and facial or soft tissue injuries [59, 60]. Concussions and other neurological injuries within the cranial vault have also been noted [61]. Snowmobiling-associated spinal injuries occur across all parts of the vertebrae-including the vertebral body, spinous process, transverse process, and facet-and at any level of the spine [61]. While rare, brachial plexus injuries have also been described [58]. Lower extremity injuries including fractures and knee dislocations in snowmobilers are often associated with vascular injuries [58].

6.4.8 Climbing

Hand and wrist injuries are particularly common in rock climbing, as climbers must exert large amounts of force through their hands [62]. Examples of climbing-specific hand and wrist injuries include hamate fractures, annular pulley system ruptures, and flexor tendon avulsions [62, 63]. Epiphyseal stress fractures have also been noted within the fingers of young rock climbers [64].

Frequent elbow conditions observed in rock climbing include climber's elbow (medial epicondylitis) and lateral epicondylitis [65, 66]. Overuse injuries encompass brachialis sprains or tendinitis, and nerve compression syndromes affecting the median, ulnar, and radial nerves [66]. Shoulder injuries are also common and include superior labrum anterior to posterior (SLAP) tears, coracoid impingement, and rotator cuff strains or tears [62, 66, 67]. Notable climbing-specific injuries are "climber's back," characterized by an increased kyphosis angle in the erect posture and, more rarely, osteochondritis dissecans in the talus and proximal interphalangeal finger joints [68–71]. Foot and ankle injuries can result from falls, contact with objects surrounding the climber, or compression from tight-fitting shoes [72]. Chronic compression from tight-fitting climbing shoes has been associated with injuries and deformities such as hallux valgus, nail disease, recurrent ankle sprains, and soft tissue inflammation including bursitis, Achilles tendinitis, plantar fasciitis, and metatarsalgia [73–75] (Table 6.1).

Table 6.1 Unique and common injury patterns in sports at altitude

General high-altitude illnesses	Acute mountain sickness High-altitude pulmonary edema High-altitude cerebral edema Photokeratitis Severe facial sunburn Frostbite
Snowboarding	High-energy distal radius fractures secondary to fall on outstretched hand Pelvic ring fractures Lateral process of the talus fracture ("snowboarder's fracture")
Alpine skiing	Clavicle fractures Isolated or multiligamentous knee injuries, especially of the ACL and MCL Ulnar collateral ligament tear of the thumb metacarpophalangeal joint ("skier's thumb")
Cross-country skiing	Cubital tunnel syndrome/ulnar neuropathy at the elbow
Freestyle, ski jumping, Nordic combined	Vertebral fractures, particularly of the lumbar and sacral region
Snowmobiling	Polytrauma—including multiple extremity and pelvis fractures and visceral injuries Brachial plexus injuries
Climbing	Hand annular pulley injuries, especially at the A2 pulley Flexor tendon strain and avulsions Medial and lateral epicondylitis ("climber's elbow") SLAP tears Abnormally increased thoracic kyphosis (>40°) ("climber's back") Calcaneal and talus fractures Osteochondritis dissecans in the talus and proximal interphalangeal finger joint Recurrent ankle sprains Nail disease Foot/ankle soft tissue inflammation

6.5 Management

Proper management of injuries at altitude is especially important given the remote and isolated nature of the terrain, which may delay the time frame for evacuation to a comprehensive medical facility. There are several steps that can be taken to ensure that injured mountain-based athletes have access to immediate care.

6.5.1 General Management

For athletes participating in a non-organized or recreational sport at altitude, such as snowmobiling, climbing, hiking, or mountain biking, there may not be medical personnel nearby in the event of injury or illness. Therefore, it is important for athletes to come prepared with medical supplies and medical knowledge to mitigate any acute injuries or illnesses until extraction is possible. Bringing first aid kits with bandages, splinting materials, antibacterial ointment, tweezers, and other supplies is highly recommended.

Staying up to date on current weather conditions prior to any mountain sport participation is advisable so that athletes can take appropriate steps to avoid heat stroke or hypothermia. For exertional heat stroke, ice water immersion has been shown to be effective and safe. In cases of non-exertional heat stroke, evaporative plus convective cooling is a better option. This strategy can be implemented through application of ice packs specifically to the neck, axilla, and groin, use of cooling blankets, or, as a supplement, administration of chilled intravenous fluids [76].

In hypothermia, patients should be actively but gently warmed as soon as possible. The heat

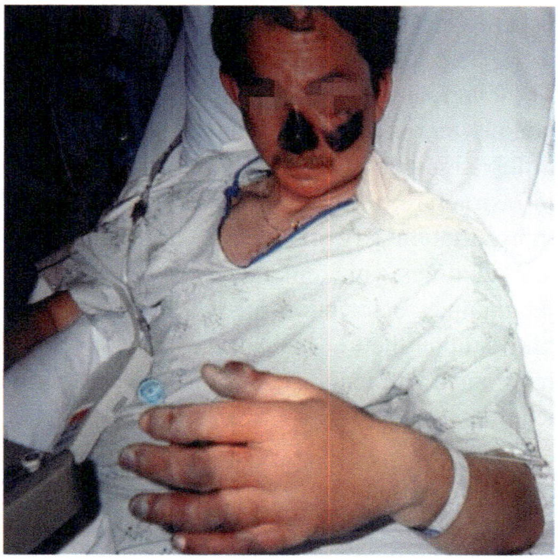

Fig. 6.2 Mountaineer in hospital with frostbitten face and hands following severe exposure while trapped on Mount McKinley in temperatures of minus 30C in 1999. (Photo Credit: Nigel Vardy (https://wellcomecollection.org/works/r7sy5672))

produced by shivering is essential for the attenuation and reversal of the hypothermia [77]. For severely hypothermic patients in whom it is difficult to confirm cardiac or respiratory activity, the patient should first be ventilated for a short period of time, after which if a pulse truly cannot be detected, CPR may be initiated [77].

For patients with frostbite, the standard of care is rapid rewarming in warm water, with temperatures from 40 to 42 °C [78]. Thrombolytic therapy has been shown to improve tissue salvage, but for severe cases with extensive ischemia or eschar formation, surgical intervention with debridement and soft tissue coverage is often necessary [78]. Surgery is performed acutely if there are signs of infection, gangrene, or compartment syndrome. Subacutely, surgical salvage should be performed before the development of dry eschar in order to promote tissue revascularization [78]. Bone scans or SPECT/CT scans can measure the extent of damage, and delayed surgery 2–4 weeks after injury—involving bone scan-guided amputation, dermal tissue grafts, local flaps, or free flaps—is recommended for reconstruction of unsalvageable tissue (Fig. 6.2) [78].

6.5.2 Skiing/Snowboarding

At any ski resort where an athlete is skiing or snowboarding on a marked ski run, any injuries or illnesses will prompt a swift response from the mountain's ski patrol unit (Fig. 6.3). It is required in the United States, Canada, and Europe for ski patrollers to be certified outdoor emergency care technicians (OEC-T) [79]. The OEC-T curriculum is more comprehensive than that for an emergency medical responder (EMR) but less than that of an Emergency Medical Technician (EMT) [80]. The main distinguishing feature is that in addition to basic emergency medical care, OEC-Ts are specially trained the treatment of environmental medical issues, such as altitude illness, hyperthermia, and hypothermia [80].

Alpine ski patrollers are trained to evacuate injured patients from the slopes by stabilizing them in the field, loading them onto a toboggan, and steering the toboggan downhill to a first aid room or to a waiting ambulance [81]. In the field, alpine ski patrollers must be able to apply bandages and splints and backboard patients with suspected spinal injuries, despite the grade of the trail or imminent weather conditions [81]. For skiers with ankle injuries, ski patrollers will often recommend that patients leave their boots on until they can access definitive treatment. The rigidity of ski boots allows effective immobilization of the injured area.

Fig. 6.3 Mount Mansfield Ski Patrol top shack in Stowe, Vermont. (Photo Credit: Victoria Bergstein)

In addition to the critical skills utilized by alpine ski patrollers, cross-country or Nordic ski patrollers are trained to improvise emergency shelters in case they need to bivouac with a patient. Cross-country ski patrols are often partnered with the US Forest Service in the United States [81].

However, it is important to note these services are only guaranteed to athletes who ski or snowboard on recognized trails. For those adventuring out of bounds, ski patrol units often cannot safely or effectively access unmarked trails to evacuate a patient [81].

6.5.3 High-Altitude Illness

For athletes suffering from AMS or HAI while at extreme altitudes, the most pivotal intervention is *rapid descent* [11]. Potentially fatal conditions like HAPE and HACE are primarily mediated by the high pressures and low oxygen content of the air at high altitudes. If caught early, those effects can be reversed with a return to a lower altitude. Besides descent, other crucial non-pharmacologic interventions include administration of supplemental oxygen, and submersion in portable hyperbaric chambers—also known as Gamow bags—as a temporizing measure, in the case of delayed evacuation [11].

In terms of pharmacologic treatment [82], HAPE is commonly treated with inhaled salmeterol, which increases alveolar fluid clearance, tadalafil and nifedipine, which both act as pulmonary vasodilators, and dexamethasone, which stabilizes the capillary endothelium [83]. Dexamethasone is also the gold standard medication for HACE treatment [82]. Importantly, acetazolamide is crucial for HACE prevention but is not efficacious for its acute treatment [82]. For HAPE, HACE, and all other AMS illnesses, prevention is the most key intervention, which can be achieved through gradual ascent with time for acclimatization [83].

6.6 Extraction Challenges

The extraction of injured athletes from mountainous terrain presents unique challenges that differ significantly from settings with more infrastructure, such as areas with well-developed transportation networks, readily accessible medical facilities, and emergency services. On a mountain, the selection of extraction resources hinges upon the injured participant's location in relation to the medical response or evacuation team.

The rapid response capabilities of helicopter rescue can significantly enhance extrication efficiency, particularly for critically ill patients or those in imminent danger. However, reductions in air density and extreme temperatures impact a helicopter's ability to generate lift and may limit engine performance [84]. Nuanced challenges associated with landing the helicopter or executing hovering operations in uneven and elevated environments further complicate matters. The most restrictive landing pads for helicopters require a level surface measuring 15 feet by 15 feet and must be free of obstructions such as trees, stumps, large rocks, or anything over 18 inches [84]. These standards are difficult to meet in many mountain settings; therefore, alternative extraction strategies may often be needed [84] (Fig. 6.4).

Ground-based rescue techniques, including the utilization of ambulances, complement airborne operations for the transportation of injured athletes [85]. These techniques offer versatile options for extracting injured athletes, particularly in situations where helicopter access is limited or impractical. The deployment of safety equipment alongside skilled personnel trained in navigating treacherous mountain landscapes aids both the safety and well-being of rescuers and patients [85].

Despite the advantages of helicopter and ground-based operations, challenges persist in high-altitude extrication scenarios, including

Fig. 6.4 A helicopter finds a level landing surface to extract injured skiers from a mountain. (Photo credit: Victoria Bergstein)

reduced supplies and resources, and limited monitoring capabilities. Medical helicopter rescue crews are advised to have at minimum technical alpine skills, including advanced expertise in rock climbing, ice navigation, ropework, belaying techniques, and alpine hazard knowledge [86]. The International Climbing and Mountaineering Federation's Medical Commission has proposed standardized protocols for assessing helicopter rescue personnel, addressing the unique demands and risks of high-altitude operations, including factors like physical fitness, altitude challenges, and occupational hazards [87]. The International Commission for Mountain Emergency Medicine has recommended multiple operational infrastructure, safety and coordination, and resource management for helicopter rescue [88].

There is also an inherent unpredictability of mountainous environments. Microexplosive charges have been used in rescue operations to overcome obstacles such as heavy rock walls, facilitating patient extraction [89]. Heightened awareness among rescuers and meticulous planning are imperative to navigate these challenges effectively and ensure the successful extrication of injured athletes.

6.7 Prevention

Although high-altitude sports will always carry an element of risk, effective patient education is a primary strategy for preventing and limiting injury. A comprehensive understanding of both general and sport-specific advice is ideal in educating athletes.

6.7.1 High-Altitude Illness Prevention

Anyone traveling to high altitudes should take appropriate precautions against HAI to prevent development of AMS, HACE, and HAPE. The most effective prevention strategies for HAI are acclimatization and gradual ascent. Guidelines from the Wilderness Medical Society recommend ascension to sleeping elevation of no more than 2750 m per day. Furthermore, at an elevation above 3000 m, they recommend ascending no more than 50 m per night and allowing an extra night of acclimatization for every 1000 m of sleeping gain [90].

For travelers and athletes needing to rapidly ascend and/or have a history of AMS, pharmacologic prophylaxis can be prescribed [1]. Prophylaxis with acetazolamide should be started the day before ascent and continued for 2–4 days after reaching the target altitude. As a carbonic anhydrase inhibitor, acetazolamide produces a metabolic acidosis that can alleviate respiratory inhibition caused by hypoxia-induced hypocapnia [82]. This effect can increase respiratory rate, which restores oxygenation and enhances the acclimatization process. Other drugs previously shown to be effective for HAI prevention include dexamethasone, which is reserved for patients who cannot take acetazolamide, and ibuprofen [82]. Nifedipine and phosphodiesterase-5 inhibitors are suitable options for HAPE prevention [91]. Some studies have also demonstrated the effectiveness of *Ginkgo biloba*—a natural supplement extracted from the ginkgo tree native to East Asia—in preventing AMS [1].

6.7.2 Climbing

Climbing injuries are common in both experienced and inexperienced climbers, but effective prevention for each group takes different forms. For beginner climbers, ensuring proper training, practice, and gear to accomplish routes is a key component of injury prevention [14]. Conversely, for advanced climbers, emphasis should be placed on carefully balancing the risks of challenging routes and climbing ability [14].

With head injuries comprising 20% of climbing injuries, a helmet may be one of the most important pieces of climbing equipment. Helmets designed specifically for alpine climbing are available and should be used regardless of climber skill. Up to 80% of climbers in certain areas do not wear helmets, highlighting the need for education on an easy-to-implement precaution that could mean the difference between safety and debilitating injury [92].

6.7.3 Mountaineering

Winter environmental hazards are a major risk factor for mountaineering, where athletes are challenged to summit mountains against the backdrop of frigid winter landscapes. Injuries often stem from the cold exposure and the consequences of lack of appropriate clothing and equipment. Weather forecasting is essential in mitigating risk for hypothermia and frostbite. Low visibility conditions may also increase the risk of falls and prevent arrival of rescuers in the event of injury. Mountain weather is dynamic and can vary significantly from initial forecasts when athletes start their ascent [93]. Thus, athletes need to be cognizant of unexpected weather and change summiting plans if weather does not permit safe ascent [94]. Athletes should also consider carrying barometers, which can indicate the possibility of an impending storm with rapid changes in air pressure [95]. Experienced local guides may also be vital in identifying safe weather conditions for ascent.

Planning for appropriate equipment and supplies is also essential to ensuring safety during mountaineering. Recently, advances in weight reduction of necessary equipment like backpacks and crampons have made it increasingly easier for mountaineers to safely pack more equipment [96]. Extra clothing can prevent frostbite in the event of unanticipated extended expeditions or a drop in temperatures while mountaineering. When planning extremely high-risk mountaineering trips, hiring a professional guide is a recommended option to ensure safety with expert advice [18].

6.7.4 Snowboarding and Skiing

Several pieces of equipment have been shown to reduce injury in skiers and snowboarders. Skiers and snowboarders have 35% lower risk of sustaining a head injury when wearing a helmet [97]. Wrist guards can be especially useful for snowboarders, among whom wrist-related injuries are highly prevalent. Wearing wrist guards has been shown to reduce the risks of wrist fracture by 71% and wrist sprain by 83% [97]. In skiers with histories of ACL injuries, functional ACL braces may help prevent further injury. Without knee braces, risk of recurrent general knee injury is 2.7 times greater among athletes recovering from ACL injuries, emphasizing the importance of injury prevention education in tandem with existing ACL injury recovery strategies [98] (Table 6.2).

Table 6.2 Prevention strategies for high-altitude sports

	Top two active prevention strategies
HAI	1. Acclimatization education: gradual ascent 2. Drug prophylaxis: acetazolamide, dexamethasone, ibuprofen, nifedipine, phosphodiesterase-5 inhibitors, and *Ginkgo biloba*
Climbing	1. Helmets 2. Risk assessment education
Mountaineering	1. Cold environment education and weather planning 2. Planning for extended expeditions with extra materials
Snowboarding and skiing	1. Equipment: wrist and knee braces when appropriate 2. Participation in snow sport safety programs

6.8 Take-Home Pearls

- All sports at altitude warrant additional caution due to decreased atmospheric oxygen content and temperature.
- Potentially fatal high-altitude-induced illnesses include acute mountain sickness, high-altitude pulmonary edema, and high-altitude cerebral edema. Gradual ascent with adequate acclimatization is key to preventing these serious altitude illnesses. Treatment of these conditions includes rapid descent of 1000–3000 feet or until symptoms resolve. Gradual ascent is key to preventing these serious altitude illnesses.
- A disproportionate number of sports injuries at altitude result from falls, which makes helmet use especially important in preventing traumatic brain injury.
- Equipment checks are vital for preventing equipment malfunction-related injuries at high-altitude, especially when climbing or mountaineering. Failure of protective gear may result in fatal falls during climbing.
- Extraction can be challenging due to the rugged and isolated nature of the mountainous terrain; therefore, effective on-site management is paramount. This includes preparation of first aid kits unique to the altitude and sport that the athlete is engaged in, as well as training on how to administer aid before emergency medical services arrive.

References

1. Hackett PH, Roach RC. High-altitude illness. N Engl J Med. 2001;345:107–14. https://doi.org/10.1056/NEJM200107123450206.
2. Mathew TM, Sharma S. High altitude oxygenation. In: StatPearls. Treasure Island (FL): StatPearls Publishing; 2024.
3. Schobersberger W, Leichtfried V, Mueck-Weymann M, Humpeler E. Austrian moderate altitude studies (AMAS): benefits of exposure to moderate altitudes (1,500–2,500 m). Sleep Breath Schlaf Atm. 2010;14:201–7. https://doi.org/10.1007/s11325-009-0286-y.
4. Beijing 2022 Audience Insights Report. n.d.
5. PyeongChang 2018 Global Broadcast and Audience Report. n.d.
6. Davidson TM, Laliotis AT. Alpine skiing injuries. A nine-year study. West J Med. 1996;164:310–4.
7. Xu Y, Yang C, Yang Y, Zhang X, Zhang S, Zhang M, et al. A narrative review of injury incidence, location, and injury factor of elite athletes in snowsport events. Front Physiol. 2021;11:589983. https://doi.org/10.3389/fphys.2020.589983.
8. Bogage J. How Mount Everest became a tourist destination. Washington Post; 2019.
9. Krakauer J. *Into Thin Air*. Penguin Random House LLC; 1997.
10. Swenson ER. High-Altitude Pulmonary Edema. In: Yuan JX-J, Garcia JGN, West JB, Hales CA, Rich S, Archer SL, editors. Textb. Pulm. Vasc. Dis. Boston: Springer US; 2011. p. 871–88. https://doi.org/10.1007/978-0-387-87429-6_61.
11. Hackett PH, Roach RC. High altitude cerebral edema. High Alt Med Biol. 2004;5:136–46. https://doi.org/10.1089/1527029041352054.
12. Smedley T, Grocott MP. Acute high-altitude illness: a clinically orientated review. Br J Pain. 2013;7:85–94. https://doi.org/10.1177/2049463713489539.
13. Schöffl V, Morrison A, Schwarz U, Schöffl I, Küpper T. Evaluation of injury and fatality risk in rock and ice climbing. Sports Med. 2010;40:657–79. https://doi.org/10.2165/11533690-000000000-00000.

14. Smith LO. Alpine climbing: injuries and illness. Phys Med Rehabil Clin N Am. 2006;17:633–44. https://doi.org/10.1016/j.pmr.2006.05.003.
15. Schussman LC, Lutz LJ, Shaw RR, Bohnn CR. The epidemiology of mountaineering and rock climbing accidents. J Wilderness Med. 1990;1:235–48. https://doi.org/10.1580/0953-9859-1.4.235.
16. Addiss DG, Baker SP. Mountaineering and rock-climbing injuries in US National Parks. Ann Emerg Med. 1989;18:975–9. https://doi.org/10.1016/S0196-0644(89)80463-9.
17. Runer A, Lampl K, Neunhäuserer D, Runer F, Frick N, Seitlinger G, et al. A 1-year prospective analysis of ice climbing injuries. Clin J Sport Med. 2017;27:161. https://doi.org/10.1097/JSM.0000000000000326.
18. Harirchi I, Arvin A, Vash J, Zafarmand V, Conway G. Frostbite: incidence and predisposing factors in mountaineers. Br J Sports Med. 2005;39:898–901. https://doi.org/10.1136/bjsm.2004.016097.
19. Campbell AD, McIntosh SE, Nyberg A, Powell AP, Schoene RB, Hackett P. Risk stratification for athletes and adventurers in high-altitude environments: recommendations for preparticipation evaluation. Wilderness Environ Med. 2015;26:30–9. https://doi.org/10.1016/j.wem.2015.09.016.
20. Ansari M, Nourian R, Khodaee M. Mountain biking injuries. Curr Sports Med Rep. 2017;16:404. https://doi.org/10.1249/JSR.0000000000000429.
21. Chow TK, Kronisch RL. Mechanisms of injury in competitive off-road bicycling. Wilderness Environ Med. 2002;13:27–30. https://doi.org/10.1580/1080-6032(2002)013[0027:MOIICO]2.0.CO;2.
22. Kronisch RL, Pfeiffer RP. Mountain biking injuries. Sports Med. 2002;32:523–37. https://doi.org/10.2165/00007256-200232080-00004.
23. Carmont MR. Mountain biking injuries: a review. Br Med Bull. 2008;85:101–12. https://doi.org/10.1093/bmb/ldn009.
24. Laver L, Pengas IP, Mei-Dan O. Injuries in extreme sports. J Orthop Surg. 2017;12:59. https://doi.org/10.1186/s13018-017-0560-9.
25. Freschi M. Epidemiology of Alpine Skiing Injuries. In: Schoenhuber H, Panzeri A, Porcelli S, editors. Alpine Ski Injuries. Cham: Springer International Publishing; 2018. p. 31–9. https://doi.org/10.1007/978-3-319-61355-0_4.
26. Leach RE, Lower G. Ankle Injuries in Skiing. Clin Orthop. 1985;198:127.
27. Rust DA, Gilmore CJ, Treme G. Injury patterns at a large Western United States Ski Resort with and without snowboarders: the Taos experience. Am J Sports Med. 2013;41:652–6. https://doi.org/10.1177/0363546512472045.
28. Basnyat B, Murdoch DR. High-altitude illness. Lancet. 2003;361:1967–74. https://doi.org/10.1016/S0140-6736(03)13591-X.
29. Roach RC, Hackett PH, Oelz O, Bärtsch P, Luks AM, MacInnis MJ, et al. The 2018 Lake Louise acute mountain sickness score. High Alt Med Biol. 2018;19:4–6. https://doi.org/10.1089/ham.2017.0164.
30. Paralikar SJ. High altitude pulmonary edema-clinical features, pathophysiology, prevention and treatment. Indian J Occup Environ Med. 2012;16:59–62. https://doi.org/10.4103/0019-5278.107066.
31. Vock P, Fretz C, Franciolli M, Bärtsch P. High-altitude pulmonary edema: findings at high-altitude chest radiography and physical examination. Radiology. 1989;170:661–6. https://doi.org/10.1148/radiology.170.3.2916019.
32. Oelz O, Maggiorini M, Ritter M, Waber U, Jenni R, Vock P, et al. Nifedipine for high altitude pulmonary oedema. Lancet Lond Engl. 1989;2:1241–4. https://doi.org/10.1016/s0140-6736(89)91851-5.
33. Li Y, Zhang Y, Zhang Y. Research advances in pathogenesis and prophylactic measures of acute high altitude illness. Respir Med. 2018;145:145–52. https://doi.org/10.1016/j.rmed.2018.11.004.
34. Blumthaler M, Ambach W, Rehwald W. Solar UV-A and UV-B radiation fluxes at two Alpine stations at different altitudes. Theor Appl Climatol. 1992;46:39–44. https://doi.org/10.1007/BF00866446.
35. Meier MM, Copeland K, Klöble KEJ, Matthiä D, Plettenberg MC, Schennetten K, et al. Radiation in the atmosphere—a hazard to aviation safety? Atmosphere. 2020;11:1358. https://doi.org/10.3390/atmos11121358.
36. Warren SG. Optical properties of ice and snow. Philos Transact A Math Phys Eng Sci. 2019;377:20180161. https://doi.org/10.1098/rsta.2018.0161.
37. Ogawa H, Sumi H, Sumi Y, Shimizu K. Glenohumeral dislocations in snowboarding and skiing. Injury. 2011;42:1241–7. https://doi.org/10.1016/j.injury.2010.12.004.
38. Deady LH, Salonen D. Skiing and snowboarding injuries: a review with a focus on mechanism of injury. Radiol Clin North Am. 2010;48:1113–24. https://doi.org/10.1016/j.rcl.2010.07.005.
39. Dunn KA. What are the health hazards of snowboarding? West J Med. 2001;174:128. https://doi.org/10.1136/ewjm.174.2.128.
40. Ogawa H, Sumi H, Sumi Y, Shimizu K. Pelvic fractures resulting from snowboarding. Am J Sports Med. 2010;38:538–42. https://doi.org/10.1177/0363546509348756.
41. Tucker NJ, Kelley N, Strage KE, Mauffrey C, Parry JA. Pelvic ring injuries after alpine ski and snowboard accidents. Eur J Orthop Surg Traumatol. 2023;33:1675–81. https://doi.org/10.1007/s00590-022-03331-x.
42. Boon AJ, Smith J, Zobitz ME, Amrami KM. Snowboarder's talus fracture: mechanism of injury. Am J Sports Med. 2001;29:333–8. https://doi.org/10.1177/03635465010290031401.

43. Kim S, Endres NK, Johnson RJ, Ettlinger CF, Shealy JE. Snowboarding injuries: trends over time and comparisons with alpine skiing injuries. Am J Sports Med. 2012;40:770–6. https://doi.org/10.1177/0363546511433279.
44. Shin CS, Chaudhari AM, Andriacchi TP. Valgus plus internal rotation moments increase anterior cruciate ligament strain more than either alone. Med Sci Sports Exerc. 2011;43:1484–91. https://doi.org/10.1249/mss.0b013e31820f8395.
45. Bere T, Flørenes TW, Krosshaug T, Koga H, Nordsletten L, Irving C, et al. Mechanisms of anterior cruciate ligament injury in world cup alpine skiing: a systematic video analysis of 20 cases. Am J Sports Med. 2011;39:1421–9. https://doi.org/10.1177/0363546511405147.
46. Bere T, Mok K-M, Koga H, Krosshaug T, Nordsletten L, Bahr R. Kinematics of anterior cruciate ligament ruptures in World Cup alpine skiing: 2 case reports of the slip-catch mechanism. Am J Sports Med. 2013;41:1067–73. https://doi.org/10.1177/0363546513479341.
47. Bürkner A, Simmen HP. Unterschenkelfrakturen beim alpinen Skisport—Einfluss von Skischuhen und Unfallmechanismus. Sportverletz Sportschaden. 2008;22:207–12. https://doi.org/10.1055/s-2008-1027947.
48. Bissell BT, Johnson RJ, Shafritz AB, Chase DC, Ettlinger CF. Epidemiology and risk factors of humerus fractures among skiers and snowboarders. Am J Sports Med. 2008;36:1880–8. https://doi.org/10.1177/0363546508318195.
49. Skier's thumb | The British Society for Surgery of the Hand. n.d. https://www.bssh.ac.uk/patients/conditions/32/skiers_thumb (accessed March 5, 2024).
50. Coury T, Napoli AM, Wilson M, Daniels J, Murray R, Milzman D. Injury patterns in recreational alpine skiing and snowboarding at a mountainside clinic. Wilderness Environ Med. 2013;24:417–21. https://doi.org/10.1016/j.wem.2013.07.002.
51. Matsumoto K, Miyamoto K, Sumi H, Sumi Y, Shimizu K. Upper extremity injuries in snowboarding and skiing: a comparative study. Clin J Sport Med. 2002;12:354–9. https://doi.org/10.1097/00042752-200211000-00006.
52. Smith GA. Biomechanics of crosscountry skiing. Sports Med Auckl NZ. 1990;9:273–85. https://doi.org/10.2165/00007256-199009050-00003.
53. Fulkerson JP. Transient ulnar neuropathy from Nordic skiing. Clin Orthop. 1980;153:230–1.
54. Stenseth OMR, Barli SF, Martin RK, Engebretsen L. Injuries in elite women's ski jumping: surveillance through the 2017–18 FIS World Cup season. Br J Sports Med. 2020;54:44–8. https://doi.org/10.1136/bjsports-2019-100799.
55. Flørenes TW, Heir S, Nordsletten L, Bahr R. Injuries among World Cup freestyle skiers. Br J Sports Med. 2010;44:803–8. https://doi.org/10.1136/bjsm.2009.071159.
56. Kastner T, Junge A, Weith M, Porzig F, Disch AC, Edouard P, et al. Injuries and illnesses during the 53rd FIS Nordic World Ski Championships 2021 in Oberstdorf: a prospective cohort study. Clin J Sport Med. 2023;33:e1. https://doi.org/10.1097/JSM.0000000000001108.
57. Hisdal E, Bahr R. FIS injury surveillance system 2006–2019. Oslo Sports Trauma Research Center; 2019.
58. Stewart RL, Black GB. Snowmobile trauma: 10 years' experience at Manitoba's tertiary trauma Centre. Can J Surg. 2004;47:90–4.
59. Beilman GJ, Brasel KJ, Dittrich K, Seatter S, Jacobs DM, Croston JK. Risk factors and patterns of injury in snowmobile crashes. Wilderness Environ Med. 1999;10:226–32. https://doi.org/10.1580/1080-6032(1999)010[0226:rfapoi]2.3.co;2.
60. DeCou JM, Fagerman LE, Ropele D, Uitvlugt ND, Schlatter MG, Connors RH. Snowmobile injuries and fatalities in children. J Pediatr Surg. 2003;38:784–7. https://doi.org/10.1016/jpsu.2003.50167.
61. Plog BA, Pierre CA, Srinivasan V, Srinivasan K, Petraglia AL, Huang JH. Neurologic injury in snowmobiling. Surg Neurol Int. 2014;5:87. https://doi.org/10.4103/2152-7806.134074.
62. Schöffl V, Popp D, Küpper T, Schöffl I. Injury trends in rock climbers: evaluation of a case series of 911 injuries between 2009 and 2012. Wilderness Environ Med. 2015;26:62–7. https://doi.org/10.1016/j.wem.2014.08.013.
63. Bayer T, Schweizer A. Stress fracture of the hook of the hamate as a result of intensive climbing. J Hand Surg Eur. 2009;34:276–7. https://doi.org/10.1177/1753193408099824.
64. Bayer T, Schöffl VR, Lenhart M, Herold T. Epiphyseal stress fractures of finger phalanges in adolescent climbing athletes: a 3.0-Tesla magnetic resonance imaging evaluation. Skeletal Radiol. 2013;42:1521–5. https://doi.org/10.1007/s00256-013-1694-4.
65. González-Iglesias J, Cleland JA, del Rosario G-VM, Fernández-de-las-Peñas C. Multimodal management of lateral epicondylalgia in rock climbers: a prospective case series. J Manip Physiol Ther. 2011;34:635–42. https://doi.org/10.1016/j.jmpt.2011.09.003.
66. Bollen SR. Soft tissue injury in extreme rock climbers. Br J Sports Med. 1988;22:145–7. https://doi.org/10.1136/bjsm.22.4.145.
67. Schöffl V, Schneider H, Küpper T. Coracoid impingement syndrome due to intensive rock climbing training. Wilderness Environ Med. 2011;22:126–9. https://doi.org/10.1016/j.wem.2010.12.005.
68. Schöffl V, Küpper T. Feet injuries in rock climbers. World J Orthop. 2013;4:218–28. https://doi.org/10.5312/wjo.v4.i4.218.

69. Ross KA, Seaworth CM, Smyth NA, Ling JS, Sayres SC, Kennedy JG. Talonavicular arthroscopy for osteochondral lesions: technique and case series. Foot Ankle Int. 2014;35:909–15. https://doi.org/10.1177/1071100714540887.
70. Bellot O, Leclercq C, Guerini H. Osteochondritis dissecans of the finger in a young climber. A case report. Hand Surg Rehabil. 2023;42:547–8. https://doi.org/10.1016/j.hansur.2023.08.009.
71. Förster R, Penka G, Bösl T, Schöffl VR. Climber's back--form and mobility of the thoracolumbar spine leading to postural adaptations in male high ability rock climbers. Int J Sports Med. 2009;30:53–9. https://doi.org/10.1055/s-2008-1038762.
72. Nelson NG, McKenzie LB. Rock climbing injuries treated in emergency departments in the U.S., 1990–2007. Am J Prev Med. 2009;37:195–200. https://doi.org/10.1016/j.amepre.2009.04.025.
73. Killian RB, Nishimoto GS, Page JC. Foot and ankle injuries related to rock climbing. The role of footwear. J Am Podiatr Med Assoc. 1998;88:365–74. https://doi.org/10.7547/87507315-88-8-365.
74. Buda R, Caprio FD, Bedetti L, Mosca M, Giannini S. Foot overuse diseases in rock climbing: an epidemiologic study. J Am Podiatr Med Assoc. 2013;103:113–20. https://doi.org/10.7547/1030113.
75. McHenry RD, Arnold GP, Wang W, Abboud RJ. Footwear in rock climbing: current practice. Foot Edinb Scotl. 2015;25:152–8. https://doi.org/10.1016/j.foot.2015.07.007.
76. Gaudio FG, Grissom CK. Cooling methods in heat stroke. J Emerg Med. 2016;50:607–16. https://doi.org/10.1016/j.jemermed.2015.09.014.
77. Giesbrecht GG. Emergency treatment of hypothermia. Emerg Med. 2001;13:9–16. https://doi.org/10.1046/j.1442-2026.2001.00172.x.
78. Zaramo TZ, Green JK, Janis JE. Practical review of the current management of frostbite injuries. Plast Reconstr Surg Glob Open. 2022;10:e4618. https://doi.org/10.1097/GOX.0000000000004618.
79. National Ski Patrol History. n.d. https://www.nsp.org/Web/NSPWebsite/Navigation/About-Us/History.aspx (accessed April 28, 2024).
80. Constance BB, Auerbach PS, Johe DH. Prehospital medical care and the National Ski Patrol: how does outdoor emergency care compare to traditional EMS training? Wilderness Environ Med. 2012;23:177–89. https://doi.org/10.1016/j.wem.2012.03.002.
81. Donelan S. Prehospital medical care and the National Ski Patrol. Wilderness Environ Med. 2012;23:175–6. https://doi.org/10.1016/j.wem.2012.01.001.
82. Zelmanovich R, Pierre K, Felisma P, Cole D, Goldman M, Lucke-Wold B. High altitude cerebral edema: improving treatment options. Biologics. 2022;2:81–91. https://doi.org/10.3390/biologics2010007.
83. Stream JO, Grissom CK. Update on high-altitude pulmonary edema: pathogenesis, prevention, and treatment. Wilderness Environ Med. 2008;19:293–303. https://doi.org/10.1580/07-WEME-REV-173.1.
84. Blancher M, Albasini F, Elsensohn F, Zafren K, Hölzl N, McLaughlin K, et al. Management of multi-casualty incidents in mountain rescue: evidence-based guidelines of the International Commission for Mountain Emergency Medicine (ICAR MEDCOM). High Alt Med Biol. 2018;19:131–40. https://doi.org/10.1089/ham.2017.0143.
85. Sumann G, Moens D, Brink B, Brodmann Maeder M, Greene M, Jacob M, et al. Multiple trauma management in mountain environments – a scoping review. Scand J Trauma Resusc Emerg Med. 2020;28:117. https://doi.org/10.1186/s13049-020-00790-1.
86. Küpper T, Hillebrandt D, Steffgen J, Schöffl V. Safety in alpine helicopter rescue operations—minimal requirements of alpine skills for rescue personnel. Ann Occup Hyg. 2013;57:1180–8. https://doi.org/10.1093/annhyg/met031.
87. Küpper T, Gieseler U, Morrison A, Schöffl V. Occupational aspects of alpine helicopter rescue operations: recommendation of the Medical Commission of the Union Internationale des Associations d'Alpinisme. Wilderness Environ Med. 2023;34:517–23. https://doi.org/10.1016/j.wem.2023.07.003.
88. Tomazin I, Ellerton J, Reisten O, Soteras I, Avbelj M. Medical standards for mountain rescue operations using helicopters: official consensus recommendations of the International Commission for Mountain Emergency Medicine (ICAR MEDCOM). High Alt Med Biol. 2011;12:335–41. https://doi.org/10.1089/ham.2010.1096.
89. Sette P, Carlini M, Berti D, Rucci I, Marchiori F, Toffali F, et al. Complex alpine extrication: case report of mountain and speleological rescue cooperation. Wilderness Environ Med. 2015;26:180–4. https://doi.org/10.1016/j.wem.2014.10.005.
90. Luks AM, Auerbach PS, Freer L, Grissom CK, Keyes LE, McIntosh SE, et al. Wilderness medical society clinical practice guidelines for the prevention and treatment of acute altitude illness: 2019 update. Wilderness Environ Med. 2019;30:S3–18. https://doi.org/10.1016/j.wem.2019.04.006.
91. High Elevation Travel & Altitude Illness | CDC Yellow Book 2024. n.d. https://wwwnc.cdc.gov/travel/yellowbook/2024/environmental-hazards-risks/high-elevation-travel-and-altitude-illness (accessed March 11, 2024).

92. Rugg C, Tiefenthaler L, Rauch S, Gatterer H, Paal P, Ströhle M. Rock climbing emergencies in the Austrian Alps: injury patterns, risk analysis and preventive measures. Int J Environ Res Public Health. 2020;17:7596. https://doi.org/10.3390/ijerph17207596.
93. Matthews T, Perry B, Khadka A, Sherpa TG, Shrestha D, Aryal D, et al. Weather observations reach the summit of Mount Everest. Bull Am Meteorol Soc. 2022;103:E2827–35. https://doi.org/10.1175/BAMS-D-22-0120.1.
94. Próchniak P, Próchniak A. Future-oriented coping with weather stress among mountain hikers: temperamental personality predictors and profiles. Behav Sci. 2021;11:15. https://doi.org/10.3390/bs11020015.
95. Moore GWK, Semple JL. Weather and death on Mount Everest: an analysis of the into thin air storm. Bull Am Meteorol Soc. 2006;87:465–80. https://doi.org/10.1175/BAMS-87-4-465.
96. Sumann G, Hochholzer T, Faulhaber M, Burtscher M. High-altitude mountaineering made safer. Trauma. 2015;17:4–16. https://doi.org/10.1177/1460408614531878.
97. Warda L, Yanchar N. Canadian Paediatric society, injury prevention committee. Skiing and snowboarding injury prevention. Paediatr Child Health. 2012;17:35–6. https://doi.org/10.1093/pch/17.1.35.
98. Owens BD, Nacca C, Harris AP, Feller RJ. Comprehensive review of skiing and snowboarding injuries. JAAOS – J Am Acad Orthop Surg. 2018;26:e1. https://doi.org/10.5435/JAAOS-D-16-00832.

It's Wet

7

Jonathan Kurjan, Shrina Parekh, and Jacob Luther

7.1 Introduction to Water Sports and Environmental Challenges

Water sports offer an exhilarating spectrum of activities, from the graceful strokes of swimming to the intense dynamics of rowing, kayaking, and triathlons. These aquatic pursuits immerse athletes in an ever-shifting environment that presents unique physiological and mechanical challenges, significantly heightening their risk of injury. Unlike their land-based counterparts, water sports demand repeated movements in unstable settings, which intensify the potential for musculoskeletal strain. For instance, the repetitive flexion-extension motions in rowing and canoeing put considerable stress on the lower back, shoulders, and knees. Research highlights that over 53% of rowers suffer from lower back pain during their careers, particularly in the lumbar region, due to these continuous motions [1]. Additionally, improper technique and equipment can further escalate this risk.

The physical risks in water sports are compounded by environmental challenges that athletes must contend with. Open water swimmers, for example, face a variety of hazards, including hypothermia, strong currents, and encounters with marine life, such as jellyfish stings. Cold water immersion is a particularly dangerous environmental factor, as it can lead to a rapid decline in body temperature. This poses a serious risk, as hypothermia can reduce motor function, impair cognitive abilities, and place additional cardiovascular stress on the body. A study by Tipton et al. (2017) highlights the particular risks of cold water immersion, noting that cold environments can impair physical performance, reduce neuromuscular coordination, and ultimately lead to drowning in extreme cases. For athletes involved in endurance events, such as long-distance open water swimming or triathlons, these risks are amplified, with hypothermia emerging as one of the most significant concerns. The sudden drop in body temperature, combined with physical exhaustion and the often unpredictable weather conditions encountered in open water environments, can increase the risk of fatal incidents, including cardiac events [2].

Furthermore, drowning remains one of the leading causes of death in open water sports, often exacerbated by sudden cardiac events that may go unnoticed prior to competition. Harris et al. (2017) report that a significant percentage of fatalities in triathlons occur during the swim portion of the race, particularly in cold water environments. The research found that 67% of all deaths in triathlons were caused by sudden cardiac events during the swim phase, a phenomenon

J. Kurjan (✉) · S. Parekh · J. Luther
University of Illinois at Chicago, Department of Orthopedic Surgery and Sports Medicine, Chicago, IL, USA
e-mail: jkurja2@uic.edu

M. R. Hutchinson, S. Rocha Piedade (eds.), *Sports Medicine: It's Environmental*,
https://doi.org/10.1007/978-3-031-84493-5_7

Fig. 7.1 Wavebreak Media ltd / Alamy Stock Photo

attributed to the combination of exertion, cold water immersion, and undiagnosed cardiovascular conditions. The unpredictable nature of open water environments, combined with physical exertion and adverse weather conditions, can trigger sudden cardiac arrest in athletes, even those who are well conditioned [3].

While outdoor water sports pose a unique set of risks, indoor sports are not without hazards. Indoor swimmers, although shielded from unpredictable weather and external elements, are subject to risks associated with prolonged exposure to chlorinated environments. Chlorine is widely used as a disinfectant in pools, but exposure to high levels of chlorine can cause respiratory issues, such as exercise-induced bronchoconstriction (EIB). Research indicates that competitive swimmers are more likely to develop respiratory conditions like asthma due to extended exposure to chlorinated water [4]. The constant inhalation of chlorinated air can lead to chronic respiratory problems, affecting both performance and long-term health. Moreover, chlorinated environments can exacerbate skin conditions, which is prevalent among competitive swimmers who spend prolonged periods in pool water [5, 6] (Fig. 7.1).

7.2 Case Anecdote/Challenge

One notable incident that illustrates the dangers of water sports occurred during the 2013 Escape from Alcatraz triathlon, an annual event that includes a challenging swim through the cold, turbulent waters of San Francisco Bay. The race, renowned for its difficulty, involves a 2.4-kilometer swim where water temperatures typically hover around 13 °C (55 °F). In 2013, a 46-year-old experienced triathlete tragically died during the swim portion of the race after suffering sudden cardiac arrest. Despite the swift intervention of rescue personnel stationed along the route, the athlete could not be revived [7].

The cold water likely played a significant role in triggering the cardiac event. Cold water immersion is known to induce a cold shock response, which involves a rapid increase in heart rate, blood pressure, and peripheral vasoconstriction as the body attempts to maintain core temperature. This cold shock response is a physiological reaction that helps the body combat the sudden decrease in temperature but can also place extreme strain on the cardiovascular system [2]. In endurance athletes, particularly those who are unaware of preexisting cardiovascular conditions, this additional cardiac workload can lead to potentially fatal outcomes.

Research conducted by Kim et al. (2012) supports this, highlighting the increased cardiac risk associated with cold water immersion. The study showed that cold water significantly increases cardiovascular strain, especially during periods of physical exertion. This heightened cardiac workload can lead to sudden cardiac arrest, particularly in athletes with undiagnosed heart conditions. Unfortunately, this was likely the case in the 2013 triathlon tragedy. Kim et al.'s study emphasizes the importance of cardiovascular screenings for athletes participating in cold water events, as early detection of cardiovascular conditions could potentially save lives [8].

Beyond the risk of cardiac events, cold water immersion impairs motor coordination and cognitive function, further increasing the likelihood of drowning. As the body struggles to maintain its core temperature, hypothermia can set in, leading to muscle weakness, fatigue, and confusion. Even experienced swimmers can become disoriented in cold water, making them more vulnerable to accidents. Tipton et al. (2017) highlight that the cold shock response not only affects

Fig. 7.2 Ammentorp Photography / Alamy Stock Photo

cardiovascular health but also diminishes cognitive function, making it harder for athletes to respond effectively to sudden changes in their environment. This makes prerace preparation, such as cold water acclimatization, an essential component of water sports training. Additionally, the presence of rescue personnel trained to manage cold water-related emergencies is crucial in mitigating the risks faced by athletes during open water events [2].

The 2013 Escape from Alcatraz Triathlon incident underscores the severe risks associated with open water swimming and other water sports, even for seasoned athletes. Cardiovascular screening, cold water acclimatization, and tailored emergency protocols are vital for ensuring the safety of participants in open water sports. While such tragic events are relatively rare, they serve as powerful reminders of the potential dangers these sports pose and the importance of rigorous safety measures (Fig. 7.2).

7.3 Epidemiology

Water sports athletes face a wide array of injury and illness risks, shaped by the physical demands of their sport and the environmental conditions in which they compete. Among the most injury-prone water sports is rowing, where the repetitive nature of the movement places significant stress on the musculoskeletal system, particularly in the lower back. A comprehensive review of rowing injuries by Thornton et al. (2017) found that over half of all competitive rowers suffer from lower back pain at some point in their careers, with many experiencing chronic issues due to the repetitive flexion-extension motions involved in the sport. This is especially true among elite athletes, who often train for prolonged periods and at high intensities, increasing the likelihood of musculoskeletal injuries. In addition to lower back pain, rowers are also at risk for rib stress fractures, with an incidence rate of around 16% among elite rowers. These fractures typically occur due to repetitive microtrauma on the chest wall during rowing, particularly when athletes push their bodies to the limits during intense training or competition [9, 10].

Swimmers, particularly those who engage in long-distance or open water swimming, face significant injury risks as well. Shoulder injuries, commonly referred to as "swimmer's shoulder," are prevalent among competitive swimmers due to the repetitive overhead motion required by swimming strokes. Research by Sein et al. (2010) indicates that up to 73% of competitive swimmers experience shoulder pain during their careers, with rotator cuff tendinopathy and subacromial impingement being the most frequent diagnoses. The repetitive nature of swimming, combined with high training volumes and inadequate recovery time, makes the shoulder joint particularly vulnerable to overuse injuries. Proper technique, as well as balancing training intensity with adequate rest, is essential in mitigating these injury risks [11].

One of the most significant dangers in open water sports is the risk of drowning, often associated with sudden cardiac events. Harris et al. (2017) found that 67% of all fatalities in U.S. triathlons occurred during the swim portion, with most deaths attributed to sudden cardiac arrest. The combination of physical exertion and cold water immersion places significant stress on the cardiovascular system, particularly in athletes with undiagnosed heart conditions. This finding underscores the need for cardiovascular screening in endurance athletes, especially those participating in events that involve open water swimming [3]. Furthermore, research has shown that the risk of sudden cardiac events is higher in colder water, as cold water immersion increases cardiac workload and can exacerbate preexisting cardiovascular conditions [8].

Respiratory and dermatological issues are also common in water sports, particularly for athletes who spend prolonged periods in chlorinated environments. Competitive swimmers are at an increased risk of developing asthma and other respiratory conditions due to extended exposure to chlorinated water and air. A study by Bernard et al. (2009) found that exercise-induced bronchoconstriction (EIB) is prevalent among swimmers, with the incidence of respiratory issues being higher in those who train in indoor, poorly ventilated pools. In addition to respiratory problems, chlorinated water can also lead to dermatological conditions, such as eczema and folliculitis. The prolonged exposure to chemicals in pool water can irritate the skin, particularly in athletes who train frequently in chlorinated environments [4]. Moreover, athletes who compete in open water events must be mindful of waterborne illnesses and infections. Poor water quality in lakes, rivers, and oceans increases the risk of bacterial and parasitic infections, such as swimmer's itch, which is caused by schistosome parasites. The Centers for Disease Control and Prevention (CDC) report that waterborne illnesses are a significant concern for water sports athletes, emphasizing the importance of maintaining water quality and hygiene practices to reduce the risk of infection [5] (Table 7.1).

Table 7.1 Common injuries, risk factors, and prevalence of water sports-related conditions [6]

Water sport	Common injuries/ illnesses	Risk factors	Injury/illness prevalence
Rowing	Lower back pain Rib stress fractures Muscle strains	Repetitive flexion-extension motion High training intensity	53% experience lower back pain 16% have rib fractures
Swimming (Indoor)	"Swimmer's Shoulder" Asthma (from chlorine exposure) Respiratory issues (EIB)	Repetitive overhead motion Chlorinated environments	73% of competitive swimmers experience "Swimmer's Shoulder"
Swimming (Open Water)	Hypothermia Cardiac Arrest Drowning Jellyfish stings	Cold water immersion Undiagnosed cardiovascular conditions Long distance swimming	67% of triathlon fatalities occur during the swim portion
Diving	Tendonitis Tympanic membrane perforations Pulmonary contusion	High-impact water entry Repetitive strains on tendons	Common injuries include shoulder and elbow tendonitis
Canoeing/ Kayaking	Shoulder dislocations Lower back pain	Repetitive upper-body motion Instability of the water environment	High prevalence of shoulder injuries among competitive kayakers

7.4 Unique Injuries/Illnesses/Patterns

Water sports provide a multitude of unique pathologies related to the various sports involved in these environments. When it comes to swimming, both indoor and open water, the most common injuries are musculoskeletal, specifically involving the shoulders, spine, and knees, respectively. The shoulder is most frequently injured (swimmer's shoulder) because most of the swimming propulsion comes from the upper limbs (as much as 90%) [12]. The etiology is related to the volume of training (some swimmers may swim up to 100 km per week) and the repetition of the stroke which can, over time, produce physical changes within the joints [12]. Typically, this is due to some combination of intra-articular pathology, subacromial impingement, scapular dysfunction, or overuse. Some swimmers may exhibit joint laxity and periscapular/rotator cuff imbalances [13]. Different strokes are more prone to different injuries as well. Knee injuries are most common in breaststroke swimmers and typically occur in the medial knee due to repetitive valgus loading or with eggbeater kicking in artistic swimming [14]. Breaststroke has a five-fold higher risk for knee pain compared to other strokes [15]. Up to 25% of swimmers will have spine pain within a single season which is usually related to the hyperextension maintained for a streamlined position vs repetitive flexion during butterfly stroke or diving [13]. Overuse injuries are the most common in diving such as patellar, quadriceps, Achilles, and posterior tibialis tendonitis, as well as traumatic ligament and tendon injuries due to the entry into the water in joints such as the wrist, elbow, shoulder, and ankle [13].

Diving hosts a uniquely different set of injuries that can occur during different stages: take-off, flight/midair maneuver, or entry. High dive imposes an increased risk of knee ligament injuries and lower extremity fractures due to the greater height of the diving board which leads to increased force and velocity at water entry [13]. Other non-orthopedic injuries seen in diving include tympanic membrane perforations, vestibular abnormalities (related to acceleration and rotation prior to deceleration at water entry), corneal epithelial injuries, pulmonary contusion, and concussion [13].

Open water swimming typically involves swimming greater distances than indoor (official distances up to 88 km) which makes open water swimming the aquatic discipline with the greatest risk of injury during competition [12]. These athletes have an increased risk of dehydration, hypo−/hyperthermia, and environmental threats such as sunburn, marine life (envenomation, jellyfish stings, bites), and otitis. Open water swimming is also the first discipline in the triathlon and has the most episodes of sudden death out of the disciplines. Sudden death is largely thought to be related to clinically silent cardiovascular disease [16]. Other potential causes include immersion pulmonary edema and autonomic conflict (tachycardia activated by the temperature of the water, exercise, and competition versus bradycardia activated by face immersion and breath holding leading to a fatal arrhythmia) [12].

Drowning is a unique risk when participating in aquatic sports and can be defined as "the process of experiencing respiratory impairment from submersion or immersion in liquid" [17]. Drowning can be nonfatal if the process is interrupted, and the individual is rescued, or fatal. Initial aspiration of water results in a cough reflex however continued aspiration without rescue can result in hypoxemia, apnea, and loss of consciousness. This can ultimately lead to cardiac arrest within minutes. Hypoxic cardiac arrest typically occurs as a result of bradycardia and pulseless electrical activity as opposed to ventricular fibrillation. The drowning process usually lasts seconds to minutes; however, in rare cases like rapid hypothermia, it can take up to an hour. Primary drowning is the most common and occurs due to the misjudgment of the ability to cope with the aquatic environment while participating in sport. Secondary drowning results from an injury, trauma, or sudden illness precipitating the loss of consciousness or causing the athlete to have the inability to cope with the aquatic environment.

From a respiratory standpoint, the risk of exercise-induced bronchospasm is higher due to the inhalation of airborne chlorine from pools [13]. This leads to inflammation and bronchoconstriction of the airway as well as rhinitis. There is sufficient evidence for an increased incidence of asthma and bronchial hyperresponsiveness in young competitive swimmers [18]. However there appears to be no adverse effects of asthma control in children and adolescents with stable controlled asthma [13]. SCUBA diving further poses the risk of immersion pulmonary edema (IPE) and decompression sickness (DCS). IPE is edema and flooding of the alveoli related to pulmonary capillary failure due to increased pressure with water immersion, cold-induced vasoconstriction, and exercise [19]. Athletes typically experience dyspnea, cough, hemoptysis, and hypoxemia. Decompression sickness results from gas molecules dissolving into tissue during ascent from a dive. Type 1 DCS is characterized by nitrogen gas affecting the skin, peripheral nerves, joints, connective tissue, and muscles. Type 2 DCS can result in an arterial gas embolus from alveoli rupture, central nervous system impairment, respiratory dysfunction or impairment, and circulatory collapse. If alveoli rupture due to increased pressure during a dive, there is a risk of tension pneumothorax on ascent due to expansion of trapped air in pleural cavity [13].

There are notable dermatological conditions associated with aquatic sports. Many are infectious such as swimmer's itch (caused by schistosome dermatitis), swimming pool granuloma (caused by atypical mycobacteria), diving suit dermatitis (caused by pseudomonas), seabather's eruption (caused by larva trapped in swimwear used by saltwater swimmers), bikini bottom/folliculitis (streptococcus or staphylococcus infection from tight-fitting swimwear), and molluscum contagiosum. Contact from pool decks can result in warts, pitted keratolysis, and tinea pedis (athlete's foot), and contact from pool water or equipment can result in contact dermatitis. Further, some swimmers may experience swimmer's xerosis which results from hot showers following training [12, 13].

Boating sports such as sailing, rowing, kayaking, and canoeing have their own unique injury patterns, most of which are musculoskeletal in nature, up to 53% of all reported rowing injuries involve the low back [13]. Common rowing injuries include muscle strains, sacroiliac joint pain, lumbar disc herniation, and spondylolysis. These are typically related to the repetitive and large load placed on the low back at the catch position (legs and back flexed, arms extended, seat at the front of slides, and blade entering water in squared position), ultimately reaching peak force at mid drive (legs extend and back beginning to extend, arms and shoulder fixed, leading to back extension and arms begin to flex to accelerate the blade through the water). Rib stress injuries and chest wall injuries are important to note in rowers as there is an incidence of up to 9% with ribs 5–9 being most frequently involved [20], but incidence increases up to 16% in elite rowers [13]. It has been hypothesized that the shearing forces of the serratus anterior and obliquus externus abdominis exerted on their attachment sites of the ribs are the cause of the chest wall/rib injuries [20]. Rowers also have increased risk of skin abrasions and hand blisters. Blisters form because of the repetitive friction between skin and oar handles. Sculler's knuckles are superficial abrasions on the dorsal surface of the hands due to hitting handles together, and track bites occur when the sliding seat catches the calf or buttocks on the track resulting in abrasions or "bites" [20].

7.5 Management

There are many challenges to managing injuries or illnesses in aquatic sports as athletes may need to be extracted prior to beginning any treatment. As with any environment, the first responder or provider must always ensure the scene is safe prior to attempting management of an athlete so as not to put themselves in danger as well. If an athlete is still in the water, safe evacuation must be performed first and will be discussed later in this chapter.

Safety interventions when put into action help reduce mortality associated with drowning, which is also known as the "drowning chain of survival" [21]. First is preventing drowning as up to 90% of drowning episodes are thought to be preventable [21]. This involves safety interventions such as using life jackets and having supervision such as lifeguards on duty even for elite athletes. The rest of the chain involves recognizing distress, providing floatation to prevent submersion, evacuation from water if safe, and providing care as needed [21]. Providing care involves basic life support, performing standard checks for breathing and responsiveness, and, if the athlete has suffered cardiac arrest from drowning, then performing cardiopulmonary resuscitation (CPR). It should also be understood that cardiac arrest from drowning is typically a result of lack of oxygen as water can interfere with alveolar expansion. Chest compressions alone are not advised in individuals who have drowned [21]. Therefore, ventilation with chest compressions is vital, starting with five rescue breaths, followed by the typical 30:2, compression-ventilation ratio, as initial ventilation can be more difficult to achieve when water is in the airways. In-water resuscitation is possible; however, it should only be attempted by highly trained rescuers and consist of only ventilation. If done correctly may increase the chance of a favorable outcome [22]. Once an athlete is back safely on land, it is important to alert advanced life support teams as quickly as possible. It should be noted that the presenting rhythm of cardiac arrest from drowning is most often asystole or pulseless electrical activity (PEA). Ventricular fibrillation is rare but may present if there is coronary artery disease or if norepinephrine or epinephrine has been used. Due to the effects of hypoxic cardiac arrest and hypothermia, sometimes higher subsequent doses of epinephrine may be considered when performing advanced care life support; however, this is controversial [22, 23].

When CPR is required, the risk of neurological damage is similar to other causes of cardiac arrest. However, the colder temperatures of water have shown to have a protective effect when the athlete also exhibits hypothermia. This allows for a more favorable outcome in individuals with prolonged submersion. Hypothermia decreases metabolic activity in the brain and oxygen consumption is reduced by 5% for each reduction of 1C within the range of 37–20C [22, 24].

Treatment of decompression sickness (DCS) primarily consists of recompression by hyperbaric oxygen treatment (HBOT); however, 100% oxygen may be administered until HBOT is available which helps dissolve gasses back into tissue and out of the bloodstream [13]. Chest x-ray is recommended to ensure there is no pneumothorax and prepare the diver for HBOT. If air transportation is required, evacuation by a pressurized aircraft is recommended. If an unpressurized aircraft must be used, it needs to be kept at altitudes between 300 and 1000 feet [25]. Hypovolemia should be corrected with intravenous fluids unless immediate abdominal or thoracic surgery is being considered [13, 25]. DCS can present as much as 36 hours after ascending from a dive, and these individuals may still benefit from HBOT [25]. Type 1 DCS return to dive recommendations differ based on symptoms. Usually for mild cases that resolve quickly, diving can resume in 2 days. If there are prolonged symptoms, oxygen/HBOT is required, paresthesias are present, or there is joint involvement, then diving should be avoided for 2–4 weeks. Type 2 DCS with central nervous system involvement that resolves with recompression requires a minimum of 4 weeks before diving again, or if there are persistent/residual deficits, then diving should be permanently restricted. If an arterial gas embolus occurs, then the diver should wait a minimum of 3 months [13].

Infectious rashes such as bacterial folliculitis on bikini bottom or tinea pedis should be managed by appropriate antibiotics or antifungals. Rashes from hot tubs or whirlpools may be caused by pseudomonas and are usually self-limiting within 7–10 days. Seabather's eruption is caused by larva trapped in the swimsuit and should be managed by showering without the swimsuit, and symptomatic control with ice, antihistamines, and, depending on pain, topical corticosteroids [12]. Bites from marine animals should be managed first by removing the athlete from the exposure, controlling any present bleeding, cleaning the wound, and using broad-spectrum antibiotics. For jellyfish envenomation,

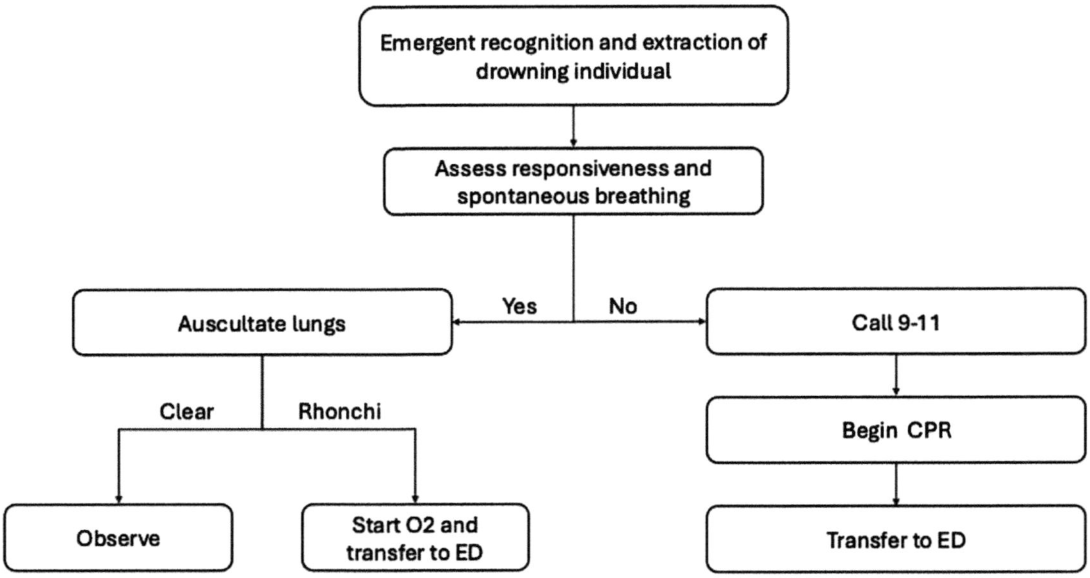

Fig. 7.3 Proposed treatment of persons who have drowned [22]

the individual should first be removed from the exposure and carefully remove any tentacles. Ice, flushing with water or vinegar, and topical anesthetics can be helpful, but for box jellyfish (sea wasp), administer vinegar to the sting site, apply oxygen and CPR as necessary, and consider treatment for anaphylaxis. Some marine animal bites may require antivenom and intensive support at a local hospital.

Musculoskeletal injuries can be evaluated on scene but ultimately can be managed appropriately in clinic or training rooms. Swimming and rowing injuries should focus on detecting improper technique, decrease training, and work on rehabilitation techniques such as mobility and strengthening [12]. For chest wall or rib stress fractures in rowers, the ribs should be offloaded, and the athlete should undergo relative rest for 3–6 weeks. Pain-free cross-training should be initiated and evaluation for relative energy deficiency in sport (RED-S), hormonal disturbances, disordered eating, and calcium/vitamin D deficiencies should be started [13] (Fig. 7.3).

7.6 Extraction

Extraction of an athlete participating in water-based activities can provide unique challenges not seen in other environments. Scene safety and ensuring that medical personnel avoid perilous situations are always priorities but should be particularly emphasized in water-based activities, which can have weather-specific and sport-specific dangers not seen elsewhere [26]. Victims who are in distress can pull rescues into the water, which could further worsen an already dangerous situation. Over a 3-year study period in Turkey, there were 114 fatalities among water rescuers [27]. Therefore, when working to extract victims of drowning or victims with injuries that do not require in-water stabilization, following the educational principle of “reach, throw, row, and go” is important to minimize the risk to responders [28]. This involves first seeing if the victim can be safely reached by extending a branch, paddle, rope, or other object while the rescuer remains out of the water on shore on in a boat. If this is not possible, the next step would be to throw the victim a flotation device such as a buoy so that they can be pulled to safety while the rescuer remains out of the water. If this is also not possible, next would be to row or paddle to the victim in a boat, kayak, canoe, or surfboard which can both support the weight of the rescuer and the victim. If choosing the "row" technique, the rescuer needs to consider whether to pull the person on board their vessel or have them hold onto the side as victims can panic and accidentally capsize a boat or surfboard [26]. Lastly, cases in which a rescuer enters the water should be considered a last result and should be utilized by professionals who have had prior experience in these rescue situations.

When working on extraction for aquatic athletes, it is important to consider the presence of other athletes or equipment that could interfere with extraction as well as the geographic location of the events. Events that take place in an indoor pool, where athletes are evenly spread out and within feet of solid ground, will have different needs than events such as triathlon or rowing events, where there will be athletes closer together, more equipment to maneuver with during extraction, and the possibility of extracting a victim further from land. It can be more challenging to treat and assess individual athletes in the water when there are other athletes, coaches, and support staff whose safety should also be considered [29]. Rowers, sailors, or triathlon swimmers may need extraction during races which can last several kilometers and whose distance from land can vary. In these situations, it can be helpful to have predesignated "extractions docks" spread along a race course to minimize the time between removing a victim from the water and finding land-based EMS [29]. Additionally, depending on the location of the sport, certain injuries may require different types of management. Cervical spine injuries which can occur more commonly in high-impact accidents such as those seen in boating, surfing, or diving may require cervical spine stabilization with surfboards, paddles, or small watercrafts, depending on their availability [28].

The water environment utilized, such as an enclosed pool, beach, pond, lake, stream, river, or open ocean, can contain its own issues that medical personnel should consider when creating an extraction plan for an athlete. Sea life such as animals and plants, rock formations, and biohazards can all create environmental hazards for victims and rescuers [26]. Strong riptides and powerful currents need to be considered when participating in sports within larger bodies of water such as lakes, beaches, open oceans, rivers, and streams. These currents can travel up to 8 feet per second, pulling even the strongest swimmers further away from shore, thereby increasing fatigue and the risk of drowning [30]. These currents account for 80% of rescues by surf beach lifeguards.

The distinction between saltwater and freshwater environments is another important distinction as 90% of drownings occur in freshwater conditions [31]. While this can be partially attributed to the pathophysiology of ingesting fresh water compared to salt water, for extraction purposes, it is important to consider the difference in buoyancy experienced between these environments. An object that displaces a cubic foot of ocean water will have an additional 1.7lbs of upward force placed on it compared with the same displacement in freshwater [32]. This amount of force can change depending on the degree of salinity in the water, temperature of the water, and the sport-specific characteristics of the victim. The higher the salt content of the water and the warmer a body of water is, the more buoyant an object will be. Additionally, whereas sailors will typically be equipped with flotation devices such as vests that increase their buoyancy, divers will commonly be equipped with weights to allow them to decrease their buoyancy in salty conditions [32]. Having an understanding of environment-specific and sport-specific buoyancy is important in assessing extraction demands and potential complications.

The potential for sudden or prolonged exposure to different water temperatures provides additional challenges for extraction. The thermal conductivity of water is 25–30 times more thermally conductive than air and any water cooler than 91.4 °F will result in heat loss [28]. Hypothermia, if untreated, can result in loss of consciousness which can be deadly in aquatic

Fig. 7.4 Caroline Cortizo / Alamy Stock Photo

environments. While the risk of hypothermia is lower in indoor swimming conditions, where water temperature is typically between 77 and 82 °F, due to the thermal conductivity of water, hypothermia injuries are still possible [33]. When assisting victims in the water, rescuers should utilize the "reach, throw, row, and go" techniques discussed earlier to minimize their own risks of hypothermia. If they need to enter cold water, rescuers should make sure to be equipped with protective equipment to prevent hypothermia such as wet suits or thermally protective dry suits [28]. When extracting hypothermia victims from open water, it is important to keep in mind that as part of the body's process of thermoregulation, removal from cold water can result in abrupt hypotension in the victim and is a known cause of post-rescue death [28]. Because of this, rescuers should maintain hypothermic patients in a horizontal position during removal from water before rewarming [28] (Fig. 7.4).

7.7 Prevention

Having proper supervision and safety protocols is paramount in preventing water-based athletic emergencies. Lifeguards and similar water-based rescuers respond to emergencies as they happen and work to prevent them from occurring. In 90% of drowning incidents, there is no supervising individual such as a lifeguard, and having a lifeguard nearby decreases the risk of beach drowning to 1 in 18 million [28, 34]. Lifeguards help participate in swimming education and drowning prevention through using verbal warnings. These prevention actions, which totaled over seven million in 2021 alone to 260 million beachgoers, help avoid potentially dangerous swimming and boating activities [34]. It is estimated that <2% of lifeguard actions involve rescues, with the remaining actions involving emergency prevention [35].

Safety personnel also help provide information on weather and water conditions which can be useful for both athletes and other rescuers. A system of flags was created and standardized in 2004 and is used to warn of potential hazards and help identify more safe swimming areas [36]. These flags contain information on safe and unsafe water conditions, but they can also denote areas where marine life is present, areas where use of watercrafts are permitted, and areas of high winds. The National Oceanic and Atmospheric Administration also helps provide information on local weather conditions that could impact aquatic sports such as head advisories, small craft advisories, special marine warnings, and thunderstorm warnings, among others [37].

Having an understanding of the type of water body used for activity and the local weather conditions can also help sports medicine physicians prevent potential exposures to waterborne bacterial or viral pathogens. As water contamination due to human feces poses a risk for athletes, having knowledge of local tidal patterns, recent storms, sources of potential contamination, and water residence time in lakes is important. Additionally, there are commercial tests which can test against adenovirus, enterovirus, and reovirus, the most common waterborne viral pathogens [38–40]. However, it should be noted that these tests can be expensive, and so, routine monitoring may be challenging. Should athletes be exposed to potentially harmful organisms in an aquatic setting, the use of prophylactic antibiotics is controversial [41]. There is a general consensus that if an athlete develops certain injuries such as a deep puncture wound or laceration, especially one on the hands, feet, face, genitals, or near a bone or joint, the administration of prophylactic antibiotics is appropriate [42].

Sports medicine physicians and athletes should also be familiar with the use of certain safety devices which can help prevent accidental drownings or other injuries. Personal flotation devices such as life jackets have been associated with 80% fewer deaths during recreational boating accidents and other studies have shown that 84% of those who died by drowning were not wearing personal flotation devices [28, 43]. Other safety equipment items can be used in certain conditions to help mitigate risk. Using wet suits or thermal protection can help protect hypothermia when swimming in cold conditions. Similarly, having access to whistles or beacon lights can help swimmers or boaters be identified in poor visibility conditions [28].

7.8 Take-Home Pearls

- *Unique Risks in Water Sports*: Water sports present distinct challenges due to unstable environments, repetitive movements, and exposure to cold or chlorinated water. Athletes face risks like hypothermia, sudden cardiac arrest, and respiratory issues from chlorine exposure, especially in competitive swimming and open water events.
- *Cold Water Hazards*: Cold water immersion can trigger a cold shock response, increasing heart rate and blood pressure, which places significant strain on the cardiovascular system. It also impairs neuromuscular coordination, raising the risk of drowning. Endurance athletes in cold water environments are especially vulnerable to hypothermia and cardiac events.
- *Common Injuries*: Rowers frequently suffer from lower back pain due to the repetitive flexion-extension movements, with over 50% experiencing pain in their careers. Swimmers, especially those engaging in long-distance swimming, often face shoulder injuries ("swimmer's shoulder") and knee injuries, particularly in breaststroke swimmers.
- *Respiratory and Skin Conditions*: Extended exposure to chlorinated environments increases the risk of respiratory issues like exercise-induced bronchoconstriction (EIB) and asthma in swimmers. Chlorinated pools can also lead to skin conditions, including eczema and folliculitis.
- *Drowning and Sudden Cardiac Events*: Drowning is a leading cause of death in open water sports, often triggered by sudden cardiac events. Cardiovascular screening and cold water acclimatization are essential preventive measures, especially for endurance athletes in events like triathlons.
- *Prevention and Management*: Lifeguards play a critical role in preventing drowning, and safety equipment such as life jackets, wet suits, and flotation devices are crucial for reducing risks. Immediate management of water-related injuries includes careful extraction, CPR, and the use of hyperbaric oxygen therapy for decompression sickness.

References

1. Bahr R, Andersen SO, Løken S, Fossan B, Hansen T, Holme I. Low back pain among endurance athletes with and without specific back loading—a cross-sectional survey of cross-country skiers, rowers, orienteerers, and nonathletic controls. Spine. 2004;29(4):449–54.
2. Tipton MJ, Eglin CM, Gennser M, et al. Immersion deaths and deterioration in swimming performance in cold water. Int J Circumpolar Health. 2017;76(1):1–8.
3. Harris KM, Creswell LL, Haas TS, et al. Death and cardiac arrest in U.S. triathlon participants, 1985 to 2016: a case series. Ann Intern Med. 2017;167(8):529–35.
4. Bernard A, Nickmilder M, Voisin C, Sardella A. Impact of chlorinated swimming pool attendance on the respiratory health of adolescents. Pediatrics. 2009;124:1110–8. https://doi.org/10.1542/peds.2009-0032.
5. Hlavsa MC, Roberts VA, Kahler AM, et al. Surveillance for waterborne disease outbreaks associated with recreational water – United States, 2009–2010. MMWR Surveill Summ. 2014;63(SS-08):1–22. Available from: https://pubmed.ncbi.nlm.nih.gov/25503921/.
6. 1.Surveillance for Waterborne Disease Outbreaks and Other Health Events Associated with Recreational Water – United States, 2007–2008 [Internet]. www.cdc.gov. Available from: https://www.cdc.gov/mmwr/preview/mmwrhtml/ss6012a1.htm.
7. Triathlete.com. Coroner investigates Alcatraz Death. Triathlete. 2020 Feb 5. Available from: https://www.

triathlete.com/culture/news/coroner-investigates-alcatraz-death/ are websites okay?
8. Kim JH, Malhotra R, Chiampas G, et al. Cardiac arrest during long-distance running races. N Engl J Med. 2012;366(2):130–40.
9. Thornton JS, Vinther A, Wilson F, et al. Rowing injuries: an updated review. Sports Med. 2017;47(4):641–61.
10. d'Ailly P, Sluiter J, Kuijer PP. Rib stress fractures among rowers: a systematic review on return to sports, risk factors and prevention. J Sports Med Phys Fitness. 2016;56:769–78.
11. Sein ML, Walton J, Linklater J, et al. Shoulder pain in elite swimmers: primarily due to swim-volume-induced supraspinatus tendinopathy. Br J Sports Med. 2010;44(2):105–13.
12. PiedadeSR NP, Espregueira-MendesJ CM, Hutchinson MR. Specific sports-related injuries. Cham: Springer; 2021.
13. Madden C, Putukian M, McCarty E, Young C. Netter's Sports Medicine. 3rd ed. Elsevier; 2022.
14. Rodeo SA. Knee pain in competitive swimming. Clin Sports Med. 1999;18(2):379–87.
15. Knobloch K, Yoon U, Kraemer R, Vogt PM. 200 – 400 m breaststroke event dominate among knee overuse injuries in elite swimming athletes. Sportverletz Sportschaden. 2008;22(4):213–9.
16. Harris KM, Creswell LL, Haas TS, Thomas T, Tung M, Isaacson E, et al. Death and cardiac arrest in U.S. triathlon participants, 1985 to 2016. Ann Intern Med. 2017;167(8):529.
17. van Beeck EF, Branche CM, Szpilman D, Modell JH, Bierens JJLM. A new definition of drowning: towards documentation and prevention of a global public health problem. Bull World Health Organ. 2005;83(11):853–6.
18. Del Giacco SR, Firinu D, Bjermer L, Carlsen KH. Exercise and asthma: an overview. Eur Clin Respir J. 2015;2(1):27984.
19. Szpilman D, Orlowski JP. Sports related to drowning. Eur Respir Rev. 2016;25(141):348–59.
20. Thornton JS, Vinther A, Wilson F, Lebrun CM, Wilkinson M, Di Ciacca SR, et al. Rowing injuries: an updated review. Sports Med. 2016;47(4):641–61.
21. Szpilman D, Webber J, Quan L, Bierens J, Morizot-Leite L, Langendorfer SJ, et al. Creating a drowning chain of survival. Theatr Res Int. 2014;85(9):1149–52.
22. Szpilman D, Bierens JJLM, Handley AJ, Orlowski JP. Drowning. N Engl J Med. 2012;366(22):2102–10.
23. Weiss SJ, Muniz A, Ernst AA, Lippton HL, Nick TG. The effect of prior hypothermia on the physiological response to norepinephrine. Resuscitation. 2000;45(3):201–7.
24. Polderman KH. Application of therapeutic hypothermia in the ICU: opportunities and pitfalls of a promising treatment modality. Part 1: Indications and evidence. Intensive Care Med. 2004;30(4):556–75.
25. Cooper JS, Hanson KC. Decompression Sickness (DCS, Bends, Caisson Disease) [Internet]. Nih.gov. StatPearls Publishing; 2019.
26. Rogers RL, Arshad FH. Chapter 14. Subspecialty: Water Rescue. In: EMRA EMS essentials: a resident's guide to prehospital care [Internet]. Dallas: Emergency Medicine Residents' Association; [cited 2024 Sept 2]. Available from: https://www.emra.org/books/emra-ems-essentials/i.-title.
27. Turgut A, Turgut T. A study on rescuer drowning and multiple drowning incidents. J Saf Res. 2012;43(2):129–32. https://doi.org/10.1016/j.jsr.2012.05.001.
28. Auerbach PS, Cushing TA, Harris NS. Auerbach's wilderness medicine. Philadelphia: Elsevier; 2017.
29. Pelto HF, Drezner JA. Design and implementation of an emergency action plan for sudden cardiac arrest in sport. J Cardiovasc Transl Res. 2020;13(3):331–8. https://doi.org/10.1007/s12265-020-09988-1.
30. Rip currents – United States lifesaving association [Internet]. [cited 2024 Sept 2]. Available from: https://www.usla.org/page/ripcurrents.
31. Juya M, Ramezani N, Peyravi G. Study of drowning in fresh and Salt Water. J Inj Violence Res. 2019;11:2. https://doi.org/10.5249/jivr.v11i2.1137.
32. Gibb N. How will your buoyancy differ in fresh vs salt water? [Internet]. TripSavvy; 2018 [cited 2024 Sept 2]. Available from: https://www.tripsavvy.com/buoyancy-salt-water-vs-fresh-water-2962936#:~:text=Salt%20water%20weighs%20more%20than%20fresh%20water%2C%20so%20it%20exerts,upward%20force%20of%2064.1%20lbs.
33. FINA Facilities Rules 2021–20254. World Aquatics; 2021 Aug [cited 2024 Sept 2]. Available from: chrome-extension://efaidnbmnnnibpcajpcglclefindmkaj/https://resources.fina.org/fina/document/2022/02/08/77c3058d-b549-4543-8524-ad51a857864e/210805-Facilities-Rules_clean.pdf#:~:text=The%20rope%20must%20effectively%20cover%20all%20lanes,Water%20temperature%20shall%20be%2025%C2%B0%20to%2028%C2%B0.
34. Davis CA, Lareau S. Drowning. Emerg Med Clin North Am. 2024;42(3):541–50. https://doi.org/10.1016/j.emc.2024.02.014.
35. Koon W, Rowhani-Rahbar A, Quan L. The ocean lifeguard drowning prevention paradigm: how and where do lifeguards intervene in the drowning process? Inj Prev. 2017;24(4):296–9. https://doi.org/10.1136/injuryprev-2017-042468.
36. Lifesaving Position Statement – LPS 14: Beach Safety and Information Flags [Internet]. Board of Directors of the International Life Saving Federation; 2010. Available from: chrome-extension://efaidnbmnnnibpcajpcglclefindmkaj/https://www.ilsf.org/wp-content/uploads/2019/01/LPS-14-2010-Flags.pdf.
37. US Department of Commerce N [Internet]. NOAA's National Weather Service; [cited 2024 Sept 6]. Available from: https://www.weather.gov/.
38. Rodríguez RA, Polston PM, Wu MJ, Wu J, Sobsey MD. An improved infectivity assay combining cell culture with real-time PCR for rapid quantification of human adenoviruses 41 and semi-quantification

of human adenovirus in sewage. Water Res. 2013;47(9):3183–91. https://doi.org/10.1016/j.watres.2013.03.022.
39. Costán-Longares A, Mocé-Llivina L, Avellón A, Jofre J, Lucena F. Occurrence and distribution of culturable enteroviruses in wastewater and surface waters of north-eastern Spain. J Appl Microbiol. 2008;105(6):1945–55. https://doi.org/10.1111/j.1365-2672.2008.03954.x.
40. Betancourt WQ, Abd-Elmaksoud S, Gerba CP. Efficiency of reovirus concentration from water with positively charged filters. Food Environ Virol. 2017;10(2):209–11. https://doi.org/10.1007/s12560-017-9332-2.
41. Noonburg GE. Management of extremity trauma and related infections occurring in the aquatic environment. J Am Acad Orthop Surg. 2005;13(4):243–53. https://doi.org/10.5435/00124635-200507000-00004.
42. Baddour LM. Soft tissue infections following water exposure. Sexton DJ, Hall KK, editors. [Internet]. 2023 [cited 2024 Sept 7]. Available from: https://www.uptodate.com/contents/soft-tissue-infections-following-water-exposure?search=water+injuries&source=search_result&selectedTitle=1~150&usage_type=default&display_rank=1#H113183489.
43. Viauroux C, Gungor A. An empirical analysis of life jacket effectiveness in recreational boating. Risk Anal. 2015;36(2):302–19. https://doi.org/10.1111/risa.12449.

8 It's Crowded (Mass Sporting Events)

Derek Altema, Sara Dykowski,
Michael Fredericson, and Seth L. Sherman

8.1 Introduction

Mass sporting events vary from endurance sports, such as running races, triathlons, and Nordic skiing, to obstacle course events and sport-specific tournament play and camps. Mass gathering events are typically characterized by the presence of at least 1000 attendees, but many mass sporting events may have many more. It can be a one-day or multiple-day event that occurs on a local, domestic, or international scale. Hundreds to thousands of participants and spectators can partake in mass sporting events, creating a unique challenge for medical providers who are tasked with ensuring the well-being of all involved. Medical providers must plan for the unexpected and start the planning process early to ensure a safe event. Environmental conditions, climate, security threats, and ill-preparedness can jeopardize the safety of participants, spectators, and staff members at mass participation events. This chapter will explore the epidemiology of injuries and illness at mass participation events, identify some unique medical challenges they present, and review strategies for management, prevention, and treatment.

D. Altema (✉) · S. Dykowski · M. Fredericson
S. L. Sherman
Stanford Health Care, Stanford, CA, USA

8.2 Case Anecdote/Challenge

You are tasked with being the medical director of your city's local 5K, half-marathon, and marathon hybrid running event that will take place in the summer (Image 8.1). You start the prerace planning 12 months in advance to coordinate with the event administrators to ensure the safety

Image 8.1 Runners begin a marathon race early in the morning

M. R. Hutchinson, S. Rocha Piedade (eds.), *Sports Medicine: It's Environmental*,
https://doi.org/10.1007/978-3-031-84493-5_8

of all participants and spectators. You find members for the medical care team committee, design protocols for medical care and rapid response as well as cancellation criteria, identify the supplies and medical equipment needs, including automated external defibrillator (AED) access, and devise an educational plan for participants and volunteers, including a heat acclimatization plan knowing the variability of the weather in July in your area. You additionally create policies and contingency plans should severe weather, terrorism events, or other crises occur, as well as policies and protocols to cover support and aid for staff and spectators. Your team confirms medical liability coverage well in advance of the event. A few months before the race, you have your medical team and race volunteers signed up. Local emergency medical services (EMS) and law enforcement are prepared to assist during the event. Your team has been educated and is ready to provide care if needed. The local trauma centers are aware of the race and prepared.

Finally, after several months of preparation, the event is here. You have monitored the weather regularly this month, and it is the same as it has been the past few weeks—a humid, warm July morning. Following safety guidelines, you review the wet bulb globe thermometer with a potential high of 21.3 °C during the race. You send out an increased risk advisory to participants for exertional heat illness, advise a slower pace for at-risk individuals, and ensure your medical team is comfortable with the exertional heat illness protocols in place before the event starts [1].

The race starts earlier in the morning to avoid the heat, and the morning is off to a smooth start. You oversee advanced medical care at one of the aid stations along the course. The 5K and half-marathoner participants finish with only a few runners with exercise-associated postural hypotension as well as minor musculoskeletal and dermatologic concerns at the finish line. The medical tent staff at the finish line triage and assist these runners cohesively as a team.

Meanwhile, it is now mid to late morning. The wet bulb globe temperature is rising, and you are working at an aid station near mile 21 in a more isolated part of the course when a runner collapses close to your aid station. Assessing the situation quickly, you assess responsiveness, as well as the participant's circulation, airway, and breathing (CABs). You recognize the signs of exertional heat illness—a rapid heartbeat, profuse sweating, and dizziness. With the help of your medical team, you move the runner to a shaded area, remove excess clothing, and begin cooling measures, including cold compresses and ice packs, as well as rehydration with electrolyte-rich fluids. A rectal temperature of 41.2 °C is taken during this time, confirming the diagnosis. You also alert the race organizers to dispatch additional medical support and arrange for transportation to the nearest medical facility for further evaluation after proper clearance. Despite the urgency of the situation, the medical team's preparedness allows you and your team to provide timely and effective cooling care, ultimately ensuring the runner's safety and well-being. After the event, you convene for a post-race debrief session to seek feedback on what went well and what could be improved upon for next year's race, constantly evaluating ways for quality improvement to further optimize participant safety.

This scenario is an example of how preparedness, timing, and education are needed to ensure the medical team is well equipped to manage medical emergencies, such as exertional heat illness, during mass participation events. Exertional heat illness is a common potential risk in endurance events, obstacle races, and sport-specific tournament play. As noted in the Current Sports Medicine Reports Team Physician 2022 Consensus Statement on mass participation and tournament event management, the necessity of excellent communication, well-defined emergency action plans with predetermined protocol and policy, including extraction, emergency personnel availability, and transport locations, equipment availability, as well as medical care staff and access during mass participation events, must be solidified before the event [2].

8.3 Epidemiology of Injury in Mass Participation Events

Understanding the epidemiology of injury and illness at mass participation events is crucial for medical personnel to adequately prepare and respond. While musculoskeletal injuries remain common across various endurance events, obstacle races, and sport-specific tournaments and camps, the specific nature and prevalence of injuries can vary significantly depending on the event type and environmental threats.

For instance, in runners, there is a high incidence of overuse injuries such as bone stress injuries, tendinopathies, and foot concerns [3, 4]. Similarly, triathlons have a high rate of overuse injuries, predominantly affecting the lower extremities, particularly related to the running and cycling segments [5–7]. The length of the course can also affect the landscape of types of medical issues arising during these distance events (Image 8.2).

A recent study examining a 3-year span in Ironman Italy found that female participants and long-distance, as opposed to Olympic/middle distance races, had a significantly higher injury risk during this specific style of event [3]. Their findings agreed with preexisting literature that systemic conditions, including muscular exhaustion, dehydration, and hypothermia, comprised the majority (68.4%) of the conditions treated during the Ironman, and only ~10% of the patients seen were for traumatic injuries, with most involving the lower limb, such as abrasions, ankle sprain, muscle strains and cramps, and plantar fasciitis [3]. However, of note, almost 10% of the participants who required evaluation by the medical team in this same study warranted hospitalization.

Image 8.2 The various types of sporting activities at mass participation events require specialized preparation and coordination of care

Additional investigation of the breakdown of medical, musculoskeletal, and dermatologic injuries during marathon and longer races vs. 10-mile events found that the percentage of medical issues increased at longer distances, while musculoskeletal and skin issues comprised 71% of issues in the 10-mile and shorter distances [1]. Therefore, at prolonged endurance events, such as an Ironman, marathons, and ultramarathons, the medical team must be prepared to triage participants and have a hardline on what can be managed in the medical tent vs. requires referral to a higher level of care. Devising a return-to-play protocol tailored to each event is essential.

In contrast to endurance events, obstacle races, such as Tough Mudder, Spartan Race, Warrior Dash, and many others, are characterized by running races with challenging terrain and obstacles. This event structure poses unique injury risks such as lacerations, sprains, strains, and blunt force trauma [8–10]. These obstacle courses are often designed for fun, teamwork-focused obstacles, such as wall climbs, fire jumps, monkey bars, barbed wire crawls, water passages, electrical wires, and many other barriers. However, with the varying degrees of athletic capability and experience in these open obstacle course events, the nature of the obstacles is a setup for potential injury. Obstacle course race events may also present environmental challenges, including exposure to extreme temperatures, which can increase the risk of heat-related illnesses or hypothermia. There are also case reports of infection related to potential contamination of water and/or ingestion of bacteria in mud or potentially close sick contacts [9].

Meanwhile, sport-specific competition, such as basketball or soccer tournaments, often involves repetitive high-intensity activities and rapid changes in direction, leading to a higher incidence of acute injuries such as ligamentous sprains and tears, muscular strains, and contusions [11]. Certain sports are also more likely to see head and neck injuries, potentially prompting the need for escalation of care with emergency department referral. Basketball, American football, and soccer were the three most common sports resulting in pediatric emergency visits for sport-related head and neck trauma, followed by lacrosse and ice hockey [12].

Sudden cardiac events may occur at any time in participants and nonparticipants alike. One case series found that over 110 sporting events, ten community cardiac arrests occurred with a breakdown of nine participants and one spectator [13]. This case series following one provider's experience over an 8-year course at mass participation events supports that out-of-hospital cardiac arrests are rare and further underlines the importance of medical preparedness as the return of spontaneous circulation (ROSC) was obtained for all individuals on site with 8/10 surviving (all shockable rhythms) and 100% of survivors making a full neurologic recovery [13].

Overall, there are multiple factors contributing to injury and illness risk, including participant demographics, event distance, environmental conditions, course terrain, and the specific demands of the sport or activity. Each participant's opportunity to acclimatize to event conditions and experience level (recreational vs. professional), as well as preemptive safety precautions, can also influence the epidemiology of injuries during mass participation events.

8.4 Unique Injuries/Illnesses/Patterns at Mass Participation Events

Mass participation events present unique medical challenges due to the large number of participants, environmental threats, and the nature of the sporting activities (see Table 8.1). In addition to the common overuse injuries seen in endurance events, certain injuries and illnesses are particularly prevalent or unique to specific event types, as previously described in the epidemiology section. Heat illness, cardiac events, musculoskeletal trauma, rhabdomyolysis, seizure, syncope, asthma exacerbations, and anaphylaxis are all potential systemic medical events that may occur to any athlete during a mass participation sporting event.

In distance events, such as running races and triathlons, as well as prolonged outdoor tournament play, exertional heat illness is a significant concern, particularly in events held in hot and humid climates, or if a team is traveling to a different climate without appropriate time for acclimation to warmer temperatures. Heat-related illnesses such as heat exhaustion and heatstroke can result from prolonged exertion combined with inadequate fluid intake and heat acclimatization [5, 14]. Hyponatremia, a potentially life-threatening condition characterized by low blood sodium levels, is another unique risk associated with endurance events, particularly in participants who overhydrate without adequate electrolyte replacement [15]. The swimming component of triathlons also raises the potential risk of drowning during the event. Prior literature notes the potential chaos at the start of the events, differences in swimmer experience level, panic, sudden cardiac events, and open vs. enclosed water can all play a role in participant drowning or near drowning [16].

Obstacle course races are associated with a higher incidence of orthopedic injuries, such as ankle sprains, knee injuries, and upper extremity injuries, such as fractures and dislocations, due to the nature of the obstacles and uneven terrain [17]. Participants in these events may also experience injuries related to falls, collisions, or overexertion while navigating obstacles. Drowning is not exclusive to swimming events and triathlons, as one participant in a Tough Mudder died in 2013 during an obstacle that involved dropping into a pool of cold, deep, murky water, further highlighting the risk involved in this style of event [16].

Table 8.1 Unique injuries of participants in mass participation events

Injury/illness	Endurance	Obstacle races	Sport specific
Exertional heat illness	Yes	Yes	Yes
Hypothermia	May occur with prolonged exposure to cold water or lower temperatures	May occur with prolonged exposure to cold water or lower temperatures	Less likely as typically a finite duration of the game
Sudden cardiac arrest	Yes	Yes	Yes
Drowning	Yes	Yes	Yes—aquatic-based sports
Rhabdomyolysis	Yes	Yes	Yes
Anaphylaxis	Yes	Yes	Yes
Exercise-associated postural hypotension (also known as exercise-associated collapse)	Yes	Yes	Yes
Acute hypoglycemia	Yes	Yes	Yes
Exercise-associated hyponatremia	Yes	Yes	Less likely, given the finite duration of play
Head and neck trauma	May occur during running or cycling events with falls or collisions. This may occur during swimming as faster swimmers can kick slower swimmers in the head when overtaking	Yes	Yes—higher risk for contact, stunt-based, and certain- ball sports
Infection	If open wounds are exposed to infected waters during triathlon or unintentional ingestion of contaminated waters	If open wounds are exposed to infected waters during triathlon or unintentional ingestion of contaminated waters	If a local wound occurs and is left untreated
Fractures	Varying degrees of bone stress injury severity given a high volume of training and intensity. It is unlikely to sustain acute traumatic fracture unless there is a fall and/or collision	Yes	Yes
Dislocations	Less likely, though, may occur with a fall or collision	Yes	Yes—sport specific
Overuse injuries	Yes—lower limb more likely	Yes—upper and lower limbs, depending on course	Yes—sport specific
Dermatologic	Yes	Yes	Yes

Meanwhile, near drowning and drowning are potential risks in aquatic sports, such as swimming and diving, artistic swim, and water polo, and it is recommended that individuals avoid swimming alone. In sport-specific mass participation events, unique injuries will be inherent to each sport as one can surmise the injuries from extreme sports, such as skateboarding, snowboarding, motocross, and rock climbing, will vary in profile from other sports, such as gymnastics, American football, golf, or baseball to name a few.

8.5 Management of Illness/Injury

As discussed above, physicians may provide care for a wide array of acute injuries/illnesses when covering a mass participation sporting event. The majority of these will be minor, will not prohibit participants from returning to competition, and are unlikely to result in hospital transport. We will discuss several conditions below that require prompt medical care during these events and face logistical challenges in management.

8.5.1 Sudden Cardiac Arrest

Managing acute cardiovascular events during a mass sporting event requires extensive preparation to ensure a prompt emergency response for athletes and spectators alike. Development of an emergency action plan with the event medical team, including the medical director/host physician and trained emergency personnel, is paramount to respond swiftly and appropriately [18, 19].

Early recognition of cardiac arrest by first responders can make a big difference in mortality and morbidity. First responders must call for assistance from others and clear surroundings to provide a safe environment in which to care for the patient. It is widely known that immediate bystander CPR before the arrival of emergency medical professionals increases the rates of meaningful survival in patients who have a cardiac arrest [20]. Rapidly accessible automatic external defibrillators (AEDs) can also improve outcomes in cardiac arrest [21]. Emergency action plans (EAP) should review locations of AEDs throughout the venue/course of the event to prevent any delays in treatment if one is needed.

The event medical team should be educated and equipped to respond to emergencies, including facilitating transport to a medical center when care escalation is required [22]. Logistical coordination with on-site EMS personnel should be addressed during planning stages, including the location of ambulances and accessibility of EMS personnel to expedite advanced cardiac life support measures and transport to higher levels of care [23]. Return of spontaneous circulation (ROSC) may be achieved before hospital transfer. Close monitoring during transport to ensure circulation, patency of the airway, and spontaneous breathing while mental status remains compromised.

8.5.2 Exertional Heat Illness

Exertional heat illness (EHI) is one of the serious medical emergencies that can arise during a mass participation event and involves elevated body temperature or hyperthermia. Participants may initially experience dizziness, nausea, weakness, and vital sign derangements. Exertional heat stroke (EHS) is when an individual's core body temperature exceeds 40 °C, with central nervous symptoms dysfunction and signs of end-organ damage [24, 25]. Recognition of this condition and acute treatment are of the utmost importance. Participants who develop EHS should be rapidly cooled on-site at the event, followed by transport to a medical center for escalation of care, using the "cool first, transport second" mantra.

Treatment should include moving the participant to a medical tent or other shaded area (Image 8.3), removing excess clothing, obtaining vital signs, including rectal temperature, and assessing mental status while communicating with advanced medical personnel in anticipation of eventual transport. There are several cooling methods available, from ice packs, cold towels, cool water baths, cool saline, and cool water immersion. Cool water immersion is the gold standard as it is the most efficient way to lower core body temperature [24, 26]. The individual should be immediately cooled until the rectal temperature reaches 39 °C and can then be transferred to the nearest medical facility via EMS. If cool water immersion is not available, a cooler dedicated to ice, water, and towels for cooling emergencies is an alternative, with the towels being rotated to ensure continued cooling effects.

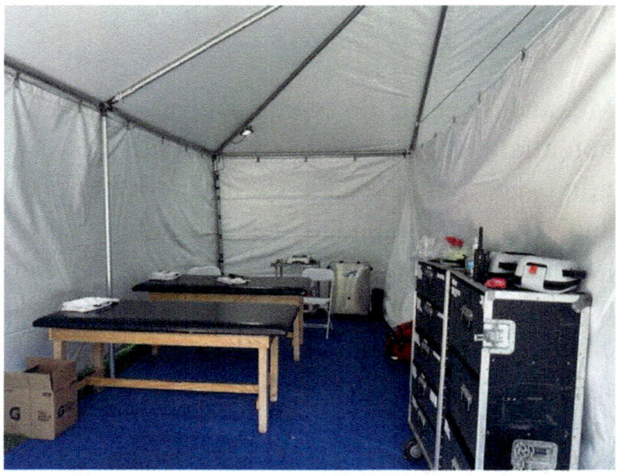

Image 8.3 An example of a medical tent that provides a shaded area to participants in need of medical attention

8.5.3 Exercise-Associated Hyponatremia (EAH)

Exercise-associated hyponatremia (EAH), defined by a serum sodium level below 135 mmol/L, is a concerning medical condition during mass sporting events, more commonly occurring in endurance athletes. There is a spectrum of pathophysiology leading to the development of EAH, classified by hypervolemia, euvolemia, or hypovolemia. In endurance athletes, the primary etiology of EAH is most often due to excessive hypotonic fluid intake/rehydration, sodium losses through sweating, and increased serum levels of antidiuretic hormone [15, 27]. Initial evaluation of fluid status, vital signs, mental status, and point-of-care serum sodium level can help to establish a diagnosis.

Symptom severity can vary widely. For those with mild to moderate hyponatremia with nonspecific symptoms (dizziness, lightheadedness, fatigue, nausea), it may be appropriate to administer oral hypertonic solutions. For patients with severe hyponatremia, characterized by headache, vomiting, or neurologic changes such as encephalopathy, seizure, or coma, the preferred treatment is intravenous hypertonic saline, typically administered in aliquots until symptoms are reversed.

Careful correction of sodium levels must be performed once a patient is transferred to a higher level of care. Rapid overcorrection of hyponatremia can precipitate a feared potential complication of osmotic demyelination syndrome [28].

8.5.4 Post-exertional Collapse

Exercise-associated collapse (EAC) is a medical condition that event physicians may encounter during mass participation in endurance events. EAC is characterized by acute-onset weakness, faintness, dizziness, or syncope, typically after an athlete reaches the finish line or completes an event [29, 30]. Exercise-associated postural hypotension (EAPH) is a form of orthostatic hypotension that occurs following a period of sustained exercise, defined by a sustained decrease in blood pressure of 20 mmHg or greater upon rising from lying supine to standing [31]. It is believed to be an underlying etiology for EAC due to the pooling of blood in the lower extremities due to impaired cardiac baroreflexes. However, it is important to avoid early diagnostic closure in athletes who suffer from post-exertional collapse; other medical conditions may also result in a collapsed athlete, such as exertional heat illness, EAH, hypoglycemia, or an acute cardiovascular event [14, 32].

A collapsed athlete should be removed from congested areas and assessed in a safe environment. After ensuring an athlete has a palpable pulse and is protecting their airway, consider ruling out other causes of collapse by evaluating mental status and core body temperature. If the underlying etiology remains unclear, it may be prudent to check serum glucose and sodium levels. When EAC is suspected, placing the body in the Trendelenburg position or elevating the legs above the level of the heart may be attempted [33]. Oral rehydration, when appropriate, may help to bolster circulating blood volume and augment recovery; IV fluids are not frequently indicated but may be considered if hypotension is sustained. Athletes who experience EAC are otherwise expected to make a full recovery [34].

8.5.5 Musculoskeletal and Traumatic Injuries

Sports medicine providers must be able to properly stabilize, immobilize, and transport participants with sports-related musculoskeletal and traumatic injuries sustained during mass participation events [35]. In the event of a traumatic incident, initial trauma ABCDE assessment should include evaluation of an athlete's airway patency, breathing, circulation, and disability/neurologic status; exposure may be addressed once safely removed from the competition area.

Some acute traumatic injuries, such as open fractures or dislocations, may require immediate care in the field with the application of an air cast, splint, or other equipment that may be readily available. Neurovascular assessment is essential to confirm the presence or absence of peripheral pulses and sensory and motor deficits. Careful examination of the affected extremity includes identifying any open wounds associated with the injury, as this may require urgent medical treatments such as irrigation, antibiotics, tetanus booster, and/or surgical fixation [36].

Prompt reduction of dislocation injuries should be performed for multiple reasons. Athletes may suffer concomitant neurovascular injury, which is likely to resolve with urgent reduction. Pain/discomfort will improve with reduction. Additionally, the longer a joint remains dislocated, the more difficult an attempted closed reduction may be and may require adjunctive medications or sedation. Contraindications to attempting immediate reduction include suspected fracture associated with the injury, complex tendinous involvement, or potential for an adverse outcome; in these cases, reduction may be deferred until imaging can be obtained [37].

Overuse injuries are common during mass participation in sporting events and are likely to be encountered by an event physician [38]. However, many overuse injuries are present before an athlete participates in an event, and while they may worsen during an event, they may not require medical evaluation during or immediately following an event unless it becomes a significant impairment [3].

8.5.6 Mass Casualty Incident (MCI)

As medical directors and host physicians of mass participation and sporting events, on-site event physicians must be familiar with disaster preparedness to manage mass casualty trauma. In the event of a mass casualty incident (MCI), the event medical team must be equipped to respond in a coordinated fashion to provide care for victims [39]. Adapting mass casualty triage algorithms for mass participation and sporting events can optimize the event medical team's ability to triage, treat, and evacuate casualties in an organized and controlled manner. This involves working with emergency medical personnel to establish efficient transport routes and distribute patients to different medical facilities and trauma centers.

Given the inherent chaotic nature of an MCI, there are many potential challenges that a medical team may encounter when managing these scenarios. There may be issues of spectator interference in crowded settings, whether due to bystander effect or attempts to volunteer [40]. Furthermore, challenges may arise due to the infrequency of MCIs; protocols that are developed in pre-event planning are rarely enacted. Communication may also be turbulent among different entities involved in patient transport [41]. Establishing effective modes of communication and a well-prepared, organized response can help to mitigate these hurdles and effectively manage these scenarios.

8.6 Extraction Challenges

A major challenge with managing illness or injury during mass sporting events is the element of safely removing participants from crowded or unsafe areas to a more controlled setting. Mass participation events take place in a wide variety of settings and environments, creating unique challenges for each particular event as it pertains to providing immediate care to an injured athlete.

Physicians need to be aware of obstacles that make patient safety and extraction difficult. Environmental factors, such as inclement weather or extreme temperatures, may create dangerous circumstances when tending to an athlete [42, 43]. Mass sporting events also present their own unique sport-specific challenges. With endurance events, medical responders must consider other nearby competitors, such as runners and cyclists on the course. Medical responders may have to navigate large crowds or physical barriers when attempting to reach a participant in need of medical care. Architectural barriers may also prevent direct access to motorized transport, e.g., paramedics may be unable to drive an ambulance directly onto the field of play to retrieve an athlete [20].

Injuries sustained during play in competitive sporting events may have extended extraction periods due to equipment constraints, such as in football and ice hockey. Trained medical personnel responding to an injured athlete may decide to safely remove protective equipment if necessary to provide emergency medical care for the patient, including situations in which airway and breathing are compromised, or there is a suspected cervical or spinal cord injury [44]. This is preferred as the first responders of the medical team that are most familiar with the equipment, such as on-field physicians and athletic trainers during football, can facilitate removal without complication. The most experienced providers should be positioned at the head of the patient to provide airway support or in-line spinal stabilization as indicated. Removal allows for access to the head and chest in case the athlete's condition worsens during transport [45].

Some race settings may be more remote, which may cause emergency medical response times to be longer [35]. Preferred methods of transport should be discussed in pre-event planning, taking into consideration the logistical challenges of the environment. Athletes who suffer serious injury or illness during events taking place in austere environments may require air transport for prompt, definitive medical treatment at a medical facility. If weather or distance impedes transfer to a hospital that can prove the definitive care, then transfer to the nearest medical facility should be facilitated for initial evaluation and stabilization [35]. On-site event physicians should be prepared to make decisions regarding methods of transport to appropriate facilities.

8.7 Prevention (Table 8.2)

8.7.1 Pre-event Planning: Emergency Action Plan (EAP)

There are many standardized requirements for the provision of medical care at mass participation events. The development of event-specific EAPs for mass participation in endurance and competition events is essential to ensure a prompt, organized care response. Pre-event planning should involve several different parties, including but not limited to medical directors and/or host physicians, event and medical support staff, emergency medical services, and event organizers [35, 46]. Working with an organized security team is important for maintaining crowd control, particularly with more densely populated events [2]. Environmental issues tend to be more prominent at mass participation sporting events, but considering variables such as weather, event type, venue setting, and expected attendance can guide pre-event planning when predicting challenges that may arise.

Developing an EAP will establish locations of emergency medical equipment, access to EMS, and clear roles within the medical response team. Ambulance or paramedic access to the competition area must also be considered in case of a medical emergency [47]. Surrounding medical facilities equipped to manage complex emergencies, such as cardiovascular events or traumatic

Table 8.2 Optimizing medical preparedness for mass participation sporting events

Pre-event planning
Develop an emergency action plan (EAP)
Coordinate with event medical team and local entities
Establish clear communication channels for event
Logistics and preparedness
Create emergency extraction protocols with event medical team and EMS
Identify nearby medical facilities, hospitals, and trauma centers for patient transfer
Consider environmental and security threats, event venue/setting, and attendance
Adapt triage protocols for managing mass casualty incidents (MCI)
Education and prevention
Encourage pre-event medical screening for athletes to stratify injury risk
Recognize common injury patterns based on type of event
Injury and illness management
Review nuances in epidemiology for mass participation sporting events
Understand management and stabilization of life-threatening injury/illness
Ensure access to early defibrillation in case of cardiovascular events

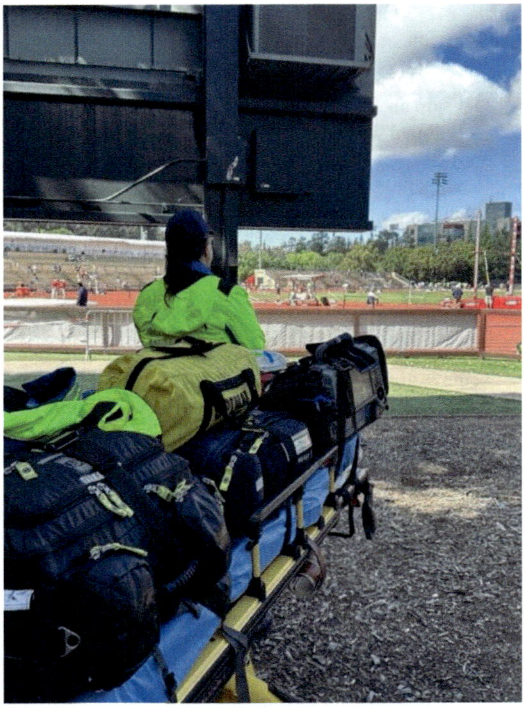

Image 8.4 A paramedic on-site, ready to provide aid should the need arise

injuries, should be identified. While the event medical team's primary responsibility is to care for the participants, planning should also discuss protocols pertaining to spectators who may develop illness. For instance, it may be difficult or impractical for an on-site host physician to tend to a spectator in the stands of a large arena. An EAP devised for a mass participation event should also include a plan to navigate a mass casualty incident (Image 8.4).

8.7.2 Logistics: Medical Support and Emergency Preparedness

Event venues vary widely in size, setting, environmental factors, and field/surface conditions [2]. Physicians providing coverage at mass participation events should work closely with the event medical team and local EMS to ensure adequate preparedness in concordance with local standards of care. The goals of the medical team include providing medical care to participants, staff, and spectators, and this may involve critical care interventions or emergency stabilization [49, 50]. As such, sports medicine physicians should be knowledgeable of common emergency conditions they may encounter and be capable of providing appropriate treatment. There should be seamless access to early defibrillation for all mass sporting events [21].

Medical directors of mass participation events must be well informed regarding local EMS and consider logistical challenges regarding transport to medical facilities. In a medical emergency, proper extraction protocols are essential for safely transporting injured or ill participants to medical facilities for further evaluation and treatment. Within an extraction protocol, a detailed outline of the course, entry/exit locations for paramedics, and the location of medical aid stations and medical staff are located should be readily available. Medical personnel should be trained in techniques for safely moving patients, including the use of stretchers, backboards, and cervical collars, in case a head or neck trauma or syncopal event occurs. Coordination with local EMS is vital to ensure timely transportation and continuity of care.

Event organizers should also implement environmental monitoring protocols to assess conditions and adjust as needed to mitigate risks. Documentation of environmental trends is recommended throughout the event to monitor the impact of environmental conditions on participant casualties [1]. During long-distance endurance events, medical tents or stations strategically located along the course allow for rapid assessment and treatment of participants. Basic medical supplies, such as bandages, gauze, wound rinses, ice packs, fluids, and fueling supplies, should be readily available. Advanced life support equipment and personnel should be on standby for more serious emergencies and purposefully distributed along the course with an appropriate ratio of medical personnel to participants. A predetermined communication system, including radios and mobile phones, to facilitate coordination between medical teams and event organizers should be in place. A protocol should be in place for how the medical team and emergency personnel will care for spectators who may sustain a serious medical issue during the event.

8.7.3 Education: Injury Prevention, Nutrition, and Fueling

Education and preparedness are at the heart of prevention plans for mitigating injury and illness before and during mass participation events. Injury patterns and frequency can vary widely depending on the event. However, not all injuries are preventable, and when they occur, effective management of injuries and illnesses at mass participation events requires a comprehensive medical plan and well-trained personnel. Education programs targeting participants, coaches, and event staff can raise awareness of risk factors and promote safety practices. An educational program should take into consideration the event type and environmental threats so that participants may train in advance to acclimatize themselves appropriately, in addition to sport-specific training and conditioning, including gradual increases in intensity and duration. Proper education can also help prevent acute in-event and overuse injuries. Hydration and nutrition guidelines tailored to the specific demands of endurance events are essential for maintaining participant health and performance.

Pre-event medical screening, including thorough musculoskeletal and overall health evaluation with cardiovascular risk assessment by a physician before the event, can help identify participants at higher risk for in-event cardiac events and other medical injuries. However, scheduling an appointment to see a medical provider is the participant's responsibility. A sports medicine physician can also meet with patients and help devise injury-preventive training to help mitigate overuse injury and acute exacerbation during in-event performance. The demanding nature of distance sports may create a mismatch in fueling and energy output, creating a low energy availability state and potential relative energy deficiency in sport (Triad/RED-S) that can predispose the athlete to bone stress injury and decreased performance, among other physiologic and psychologic negative impacts [48]. Participants who are suffering from Triad/RED-S should follow up with a specialist in low energy availability management to help facilitate safe nutrition while undergoing training to reduce risk for potential short- and long-term health consequences.

8.8 Take-Home Points

- Mass participation sporting events present unique challenges for medical providers and require diligent pre-event planning and coordination between different entities to ensure the safety and well-being of participants.
- The goals of the medical team at mass participation sporting events should include providing medical care to participants within local standards of care, utilizing emergency action plans for logistical efficiency, and maintaining safety precautions throughout the duration of the event.
- Understanding the epidemiology and management of injuries and illnesses relative to different mass participation events can optimize the medical team's response and prevent delays in definitive care.
- Considering environmental factors, security threats, and crowd congestion during the development of EAPs is essential when creating extraction and transport protocols for medical emergencies that may occur during mass participation events.

References

1. Roberts WO, Armstrong LE, Sawka MN, Yeargin SW, Heled Y, O'Connor FG. ACSM expert consensus statement on exertional heat illness: recognition, management, and return to activity. Curr Sports Med Rep. 2021;20(9):470–84.
2. Herring SA, Kibler WB, Putukian M, Blauwet C, Boyajian-O'Neill LA, Boyd J, et al. Mass participation and tournament event management for the team physician: a consensus statement (2022 update). Curr Sports Med Rep. 2024;23(4):143–58.
3. Feletti F, Saini G, Naldi S, Casadio C, Mellini L, Feliciani G, et al. Injuries in medium to long-distance

triathlon: a retrospective analysis of medical conditions treated in three editions of the ironman competition. J Sports Sci Med. 2022;21(1):58–67.
4. Van Gent RN, Siem D, Van Middelkoop M, Van Os AG, Bierma-Zeinstra SMA, Koes BW, et al. Incidence and determinants of lower extremity running injuries in long distance runners: a systematic review. Br J Sports Med. 2007;41(8):469–80.
5. Lepers R, Rüst CA, Stapley PJ, Knechtle B. Relative improvements in endurance performance with age: evidence from 25 years of Hawaii Ironman racing. Age (Dordr). 2013;35(3):953–62.
6. Jaworski CA. Medical concerns of marathons. Curr Sports Med Rep. 2005;4(3):137–43.
7. Cushman DM, Dowling N, Ehn M, Kotler DH. Triathlon considerations. Phys Med Rehabil Clin N Am. 2022;33(1):81–90.
8. Gosling CMR, Gabbe BJ, Forbes AB. Triathlon related musculoskeletal injuries: the status of injury prevention knowledge. J Sci Med Sport. 2008;11(4):396–406.
9. Rabb H, Coleby J. Hurt on the Hill: a longitudinal analysis of obstacle course racing injuries. Orthop J Sports Med. 2018;6(6):232596711877985.
10. Knapik JJ. Obstacle course events: hazards and prevention measures. J Spec Oper Med. 2022;22(2):129–38.
11. Lund A, Turris SA, McDonald R, Lewis K. Onsite management of medical encounters during obstacle adventure course participation. Curr Sports Med Rep. 2015;14(3):182–90.
12. Hawkins D, Metheny J. Overuse injuries in youth sports: biomechanical considerations. Med Sci Sports Exerc. 2001;33(10):1701–7.
13. Scheffler P, Wolter NE, Namavarian A, Propst EJ, Chan Y. Contact sport related head and neck injuries in pediatric athletes. Int J Pediatr Otorhinolaryngol. 2019;121:6–9.
14. Morton S, Goonetilleke C, Taylor M, Beach N. Out-of-hospital cardiac arrests during mass-participation endurance events: a case series. Wilderness Environ Med. 2023;34(3):318–21.
15. Hew-Butler T, Loi V, Pani A, Rosner MH. Exercise-associated hyponatremia: 2017 update. Front Med (Lausanne). 2017;4:21.
16. Eichner ER. Tough Mudder injuries, triathlon drownings, and team rhabdomyolysis in the navy. Curr Sports Med Rep. 2014;13(2):66–7.
17. Gimigliano F, Resmini G, Moretti A, Aulicino M, Gargiulo F, Gimigliano A, et al. Epidemiology of musculoskeletal injuries in adult athletes: a scoping review. Medicina (Kaunas). 2021;57(10):1118.
18. Marijon E, Tafflet M, Celermajer DS, Dumas F, Perier MC, Mustafic H, et al. Sports-related sudden death in the general population. Circulation. 2011;124(6):672–81.
19. Perkins GD, Handley AJ, Koster RW, Castrén M, Smyth MA, Olasveengen T, et al. European resuscitation council guidelines for resuscitation 2015. Resuscitation. 2015;95:81–99.
20. Serratosa L, Kramer E, Börjesson M. Cardiac safety at sports events: the medical action plan. In: Pelliccia A, Heidbuchel H, Corrado D, Börjesson M, Sharma S, editors. The ESC textbook of sports cardiology. Oxford University Press; 2019. p. 411–8.
21. Grubic N, Hill B, Phelan D, Baggish A, Dorian P, Johri AM. Bystander interventions and survival after exercise-related sudden cardiac arrest: a systematic review. Br J Sports Med. 2022;56(7):410–6.
22. Neumar RW, Shuster M, Callaway CW, Gent LM, Atkins DL, Bhanji F, et al. Part 1: executive summary: 2015 American Heart Association guidelines update for cardiopulmonary resuscitation and emergency cardiovascular care. Circulation. 2015;132(18_suppl_2).
23. Soar J, Maconochie I, Wyckoff MH, Olasveengen TM, Singletary EM, Greif R, et al. 2019 International consensus on cardiopulmonary resuscitation and emergency cardiovascular care science with treatment recommendations: summary from the basic life support; advanced life support; pediatric life support; neonatal life support; education, implementation, and teams; and first aid task forces. Circulation. 2019;140(24):e826.
24. Armstrong LE, Casa DJ, Millard-Stafford M, Moran DS, Pyne SW, Roberts WO. Exertional heat illness during training and competition. Med Sci Sports Exerc. 2007;39(3):556–72.
25. Navarro CS, Casa DJ, Belval LN, Nye NS. Exertional heat stroke. Curr Sports Med Rep. 2017;16(5):304–5.
26. Casa DJ, Armstrong LE, Kenny GP, O'Connor FG, Huggins RA. Exertional heat stroke: new concepts regarding cause and care. Curr Sports Med Rep. 2012;11(3):115–23.
27. Almond CSD, Shin AY, Fortescue EB, Mannix RC, Wypij D, Binstadt BA, et al. Hyponatremia among runners in the Boston Marathon. N Engl J Med. 2005;352(15):1550–6.
28. George JC, Zafar W, Bucaloiu ID, Chang AR. Risk factors and outcomes of rapid correction of severe hyponatremia. Clin J Am Soc Nephrol. 2018;13(7):984–92.
29. Asplund CA, O'Connor FG, Noakes TD. Exercise-associated collapse: an evidence-based review and primer for clinicians. Br J Sports Med. 2011;45(14):1157–62.
30. Roberts WO. Exercise-associated collapse care matrix in the marathon. Sports Med. 2007;37(4–5):431–3.
31. Holtzhausen LM, Noakes TD. Collapsed ultraendurance athlete: proposed mechanisms and an approach to management. Clin J Sport Med. 1997;7(4):292–301.
32. Chen AW, Archbold CS, Hutchinson M, Domb BG. Sideline management of nonmusculoskeletal injuries by the Orthopaedic Team Physician. J Am Acad Orthop Surg. 2019;27(4):e146–55.
33. Anley C, Noakes T, Collins M, Schwellnus MP. A comparison of two treatment protocols in the management of exercise-associated postural hypotension: a randomised clinical trial. Br J Sports Med. 2011;45(14):1113–8.

34. Irelan MC, Schroeder JD. Exercise-associated collapse. In: StatPearls. Treasure Island: StatPearls Publishing; 2024.
35. Khodaee M, Waterbrook AL, Gammons M, editors. Sports-related fractures, dislocations and trauma: advanced on and off-field management. Cham: Springer; 2020. 1020 p.
36. Menkes JS. Initial evaluation and management of orthopedic injuries. In: Tintinalli J, Kelen G, Stapczynski J, editors. Emergency medicine: a comprehensive study guide; 2000. p. 6.
37. Wright CJ, Brandon BA, Reisman EJ. Closed-reduction techniques for glenohumeral-, patellofemoral-, and interphalangeal-joint dislocations. J Athl Train. 2020;55(8):757.
38. Kipps C, Smith R, Knight N, Parkin J. Injuries and illness during mass-participation triathlon races: optimising competitor safety. Br J Sports Med. 2017;51(4):343.2–44.
39. Sedgley M, Hudson K, Hulsopple C. Prepare for the unexpected: a new look at trauma triage and care in mass participation sporting events. Curr Sports Med Rep. 2023;22(1):4–9.
40. Saadatmand V, Ahmadi Marzaleh M, Abbasi HR, Peyravi MR, Shokrpour N. Emergency medical services preparedness in mass casualty incidents: a qualitative study. Health Sci Rep. 2023;6(10):e1629.
41. Kohn SC, De Visser E, Weltman G, Bosson N, Gausche-Hill M, Donaldson R. Emergency medical team coordination in mass casualty incidents: a study of the Los Angeles emergency medical system. Proc Hum Factors Ergon Soc Annu Meet. 2016;60(1):523–7.
42. Rocha Piedade S, Hutchinson MR, Parker D, Espregueira-Mendes J, Neyret P, editors. Sideline Management in Sports. Cham: Springer International Publishing; 2023.
43. Vasquez MS, Fong MK, Patel LJ, Kurose B, Tierney J, Gardner I, et al. Medical planning for very large events: special olympics world games Los Angeles 2015. Curr Sports Med Rep. 2015;14(3):161–4.
44. Diduch BK, Hudson K, Resch JE, Shen F, Broshek DK, Brady W, et al. Treatment of head and neck injuries in the helmeted athlete. JBJS Rev. 2016;4(3):e2.
45. Mills BM, Conrick KM, Anderson S, Bailes J, Boden BP, Conway D, et al. Consensus recommendations on the prehospital care of the injured athlete with a suspected catastrophic cervical spine injury. J Athl Train. 2020;55(6):563–72.
46. Chiampas G, Jaworski CA. Preparing for the surge: perspectives on marathon medical preparedness. Curr Sports Med Rep. 2009;8(3):131–5.
47. Aljohani M, Kennedy KE. Mass gatherings. In: Ciottone's disaster medicine. Elsevier; 2016. p. 968–74.
48. Mountjoy M, Ackerman KE, Bailey DM, Burke LM, Constantini N, Hackney AC, et al. 2023 International Olympic Committee's (IOC) consensus statement on Relative Energy Deficiency in Sport (REDs). Br J Sports Med. 2023;57(17):1073–98.
49. Margolis AM, Leung AK, Friedman MS, McMullen SP, Guyette FX, Woltman N. Position statement: mass gathering medical care. Prehosp Emerg Care. 2021;25(4):593–5.
50. Schwartz B, Nafziger S, Milsten A, Luk J, Yancey A. Mass gathering medical care: resource document for the national association of EMS physicians position statement. Prehosp Emerg Care. 2015;19(4):559–68.

It's Dangerous

9

Matthew A. Siegel, Diego Barragan Echenique, Mario E. Ramirez, Peter Cirrincione, Alfonso Mejia, and Mark R. Hutchinson

9.1 Introduction

The beneficial impact of sports on participants, spectators, and society is myriad and well documented. Organized sports have been an integral part of civilization from the very start. Archeologic evidence from ancient Sumer from around 3000 BC suggests ancient Sumerians engaged in wrestling and boxing. Illustrations and writings from ancient Egypt from around 2600–2400 BC depict archery, wrestling, and rowing. The tomb of Beni Hasan dating from around 2000 BC contains paintings depicting wrestling matches. The tomb of Khnumhotep depicts wrestling as well as the much more dangerous staff fighting.

The Olympic Games began in Greece in 776 BC with events such as running, long jump, shot putt, javelin, boxing, and equestrian events. Unfortunately, coupled with accounts of these early events, we also see the onset of violence surrounding sporting events. In AD 59 at a gladiator show in the Amphitheater of Pompeii, a riot occurred between spectators from Pompeii and rival spectators from Nuceria. What began as typical jeering and taunting between rival townspeople quickly turned violent and resulted in a riot involving weapons and casualties. The shocked Roman Senate launched an investigation into this event that led to a 10-year ban on hosting gladiator shows at Pompeii as a measure to prevent further violence and maintain public order. Figure 9.1 depicts the gladiators circa this time.

Fig. 9.1 Bas-relief depicting pompa triumphalis/opening ceremony with gladiator and animal fights at Pompeii, Italy, 20–50 AD. (Reproduced from Alamy at https://www.alamy.com/stock-photo-bas-relief-depicting-pompa-triumphalis-opening-ceremony-with-gladiator-97257555.html?imageid=27007F97-2C3B-4FCC-9840-3FBFDCF1D381&p=80871&pn=1&searchId=bfa518094a05b6b8f882fb0eab91ac41&searchtype=0; "Photographer: Clement Philippe/Contributor: Arterra Picture Library/Alamy Stock Photo," Image ID: FJ6D03, Date taken: 27 February 2016. All rights reserved)

M. A. Siegel (✉) · D. B. Echenique · M. E. Ramirez
P. Cirrincione · A. Mejia
Department of Orthopaedic Surgery, University of Illinois at Chicago, Chicago, IL, USA
e-mail: msiegel2@uic.edu

M. R. Hutchinson
UIC Sports Medicine Center, Department of Orthopaedics, University of Illinois at Chicago, Chicago, IL, USA
e-mail: mhutch@uic.edu

M. R. Hutchinson, S. Rocha Piedade (eds.), *Sports Medicine: It's Environmental*,
https://doi.org/10.1007/978-3-031-84493-5_9

Despite this measure, this was not the last major loss of life at a Roman sporting event as documented by the riots at Nika, during a chariot race in the Hippodrome of Constantinople in 532 AD. This stemmed from the arrest of fans from the rival Blue and Green factions for murder from deaths connected with rioting after a chariot race the prior year. Indeed, reports of spectator and athlete violence at or in association with sporting events continue to this day.

Violence in sports is a form of interpersonal violence, and it can be defined as any physical or psychological incident that creates harm or has the potential to cause harm to an athlete, spectator, or other individuals involved in sporting events. This includes both physical aggression (e.g., fights, brawls) and psychological aggression (e.g., verbal abuse, harassment) [1]. Violence in sports may be intentional, such as physical assaults involving fighting between players, spectators, or a combination. Intentional violence may also involve retaliatory actions, hazing, or verbal abuse. Unintentional violence encompasses accidental collisions, excessive aggressiveness, or technical fouls. Except for excessive aggressiveness during play, unintentional violence is beyond the scope of this chapter.

Intentional violence in sports not only affects the immediate participants but also has broader implications for the culture and perception of sports. Physical assaults can lead to severe injuries and long-term health issues. Retaliatory actions create a cycle of violence that can escalate and disrupt the integrity of the sport. Hazing practices, often rationalized as team-building activities, can result in psychological trauma and physical harm, undermining team cohesion and individual well-being. Verbal abuse, while sometimes dismissed as mere "trash talk," can have significant psychological impacts on targets and contribute to a hostile environment. Addressing these issues requires comprehensive policies, education, and a call to creating a culture of respect and sportsmanship.

This chapter will discuss types of violence in sports including on-field violence, off-field violence, and psychological and verbal violence. It will outline common injuries resulting from violence both acute and chronic as well as cumulative injuries, such as concussions. The team physician must be thoughtful in his or her preparation for violence at sporting events by analyzing historical outcomes and being able to help enforce evolving protective measures.

9.2 Case Challenge

You are the event physician for an inaugural half marathon located in a highly populated city. You have set up aid stations along the course and a large medical station near the finish line. As the race is getting started, an explosion occurs outside of the medical tent. What is your response and what pre-event planning could have been optimized?

Has anyone been hurt? Do you have access to provide appropriate medical care? Is there potential danger for an additional explosion? Is there an adequate chain of command for addressing this situation? Do you cancel the entire event? Was the medical tent protected and placed in an appropriate location? Was the course surveyed and thoughtfully organized? Was there an appropriate number of medical tents set up?

Planning for an event of this magnitude should have started at least 3–12 months prior. This includes developing a disaster plan, medical protocols, supply and staffing availabilities, and many other arrangements. Law enforcement, emergency departments and hospitals, as well as emergency transportation should have been included in this planning [2]. It is also vital that all preparatory materials were provided to the necessary ancillary staff and physicians covering this event. This way, there will be a team of educated providers who are able to act on the protocols to help in this situation. Do you have your lightning and storm protocols ready in the case of inclement weather? Does this weather protocol affect your response to the explosion? You do not want to be thinking of these questions for the first time when chaos is occurring and emotions are running high.

9.3 Epidemiology, Types of Violence, and Its Effects

Violence in sports encompasses a range of behaviors that can cause physical or psychological harm to athletes, officials, and spectators. This includes intentional physical aggression, such as fighting and rough play, as well as verbal abuse and threatening behavior. The definition of violence in sports can vary depending on the context, but it generally refers to actions that exceed the accepted rules and norms of the sport. The scope of violence includes player-on-player violence, spectator violence, structural violence, and psychological violence as well as myriad combinations of these components.

In the International Olympic Committee Consensus Statement, they defined psychological abuse as the "core" of all other forms of abuse: "A pattern of deliberate, prolonged, repeated non-contact behaviours within a power differentiated relationship" [3]. This may include verbal abuse of intimidation, threats, harassment, and belittling. This form of harassment may continue, or start, as cyberbullying with online threats and insults. Psychological abuse can also be seen in the form of the neglect of proper coaching or lack of supplying needed equipment. Neglect can be defined as the failure of a coach to meet its athletes' physical and emotional needs or to protect them from harm [3]. Psychological violence is often referred to as the gateway to other forms of violence [3]. Physical violence includes fights and rough play; however, this type of violence can also be seen with overtraining that may occur from punishing an athlete's poor performance. Other forms of physical violence from coaches can be from limiting fluids and fuel as well as aggressive physical corrections of athletic form [4]. Sexual violence is defined as any sexual act when consent is not or cannot be provided [3]. This encompasses more than contact or penetrative actions, such as indecent exposure and pornographic materials [4]. These forms of sports-related violence may in part take place due to systematic and organizational abuses, which may involve deep-rooted sports mentalities and minimally acknowledging athletes' complaints. The governing organizations may provide a lackluster system that does not have accountability [4, 5].

Violence in sports is a multifaceted issue that affects various levels of competition, from youth leagues to professional sports. It is well known that many children receive a boost in their self-esteem through their participation in sports. Sports participation can also help prevent suicide and teach valuable life lessons [6, 7]. However, these positive adaptations are diminished when an athlete experiences abuse, such as psychological, physical, sexual, and/or neglect [3]. Further, many of these events that occur in youth sports may qualify as an adverse childhood experience, which is significantly correlated with worse adult health outcomes. When comparing people who have zero adverse childhood experiences to people who have one to more than four adverse childhood experiences, there is a reduction of 14.9% for unemployment, 23.9% for heavy drinking, 27.0% for chronic obstructive pulmonary disease, and 44.1% for depression [8]. These morbidities lead to attributed annual costs of over $550 billion in Europe and almost $750 billion in North America [9]. A study consisting of over 1000, 14 to 17-year-olds highlighted that 52% of the athletes experienced psychological violence, while 10% of the athletes were "poly-victimized" by their teammates, coaches, and parents [10].

These figures in the pediatric population are similar to adults competing at an elite level [11]. 57.8% of 473 current or former US Olympic athletes above the age of 18 stated that they experienced at least one type of sports-related violence. Emotional abuse was the main sports-related violence reported in this study, which was significantly linked to having a psychiatric condition such as anxiety, depression, post-traumatic stress disorder, and/or self-reported eating disorders [11]. No matter what level of athletics, the team physician must be aware of acute and chronic environments that permit this type of behavior. It is prudent to care for and potentially intervene during violent scenarios. These actions may not only diffuse a dangerous situation but also prevent the athletes from developing adult psychiatric and medical conditions.

Perpetrators and victims of violence in sports can be anyone: male and female coaches, other athletes and teammates, as well as other staff involved such as athletic or medical staff, referees, and other adjacent workers. One survey study based out of Belgium and the Netherlands detailed the perpetrators of violence that 1785 athletes experienced before they were 18 years old [12]. Their main conclusions were that the main perpetrators of psychological, physical, or sexual violence in sports were peer athletes. Further, 54–70% of these athletes reported more than one perpetrator. And if more than one perpetrator was involved in a given event, this was typically scored as significantly more severe. Additionally, the group also found that most of the perpetrators described in the study were male; however, the authors did comment that the coaching staff in this area was approximately 70% male [12]. It is vital to be aware of a typical perpetrator profile concerning a specific type of sports violence. For example, the number one reported perpetrator for psychological violence against female victims was several female athletes [12].

It is also vital to be aware of common risk factors for who potential victims can be. A study involving French Canadians who participated in all levels of athletics demonstrated that victims of multiple episodes violence were more likely to have an earlier sports specialization, to practice isolated from their parents, and to demonstrate unhealthy athletic behaviors such as extreme weight control, non-suicidal self-injury, and practicing despite injuries. There was no significant difference between team and individual sports [10]. The profiles of victims and perpetrators in the above studies may not be generalizable to all populations. The team physician must be aware of different ethnic, team, or organizational cultures in order to appropriately intercede. Again, it is important to emphasize that anyone can be a perpetrator, a victim, or both. Also, traumatic experiences lend themselves to causing more destruction and violence downstream, which can be in the form of retaliation or abuse propagation. There may be an aspect of loyalty to teammates and going after the other team. This is evident in professional baseball when the pitcher beans a batter on the opposing team after a preliminary event of misconduct.

9.4 Violence in Amateur Sports

The causes of violence in amateur sports have been studied for decades. Insight from the fields of psychology, sociology, and anthropology have each informed current literature; however, there is a dearth of research focusing on an interdisciplinary approach. Most recently, Spaaij and Schaillee have implemented a framework that analyzes the interactions of individual factors, situational factors, and social environment on aggression and violence in sports [13]. Personal values and demographic factors largely influence the individual component, whereas the personal values of desire for self-enhancement in sports and parental pressure have shown associations with excessive aggression or violent behavior in sports [14]. Male gender, older age, and higher level of competition have also been associated with increased aggression and violence in sports [15]. The team physician should be aware of these internal and external motivating forces on an individual's personal values and how these may influence an athlete's excessive aggression on the field. Specifically, it is important to recognize young athletes subjected to increased pressure for performance from coaches and/or parents and serve as an actor who encourages prioritizing physical and mental well-being to achieve increased levels of performance. Referrals to other medical providers, such as therapists or sports psychologists, may be of great value.

Situational factors refer to the type and competition level of the sport. Contact-sports athletes have been observed to engage in more aggression in sports compared to noncontact-sports athletes [13, 16, 17]. However, in this context, previous research delineates the importance of qualifying the type of excessive aggression observed as either unsanctioned, instrumental/non-provoking, or hostile/reactive. Here, increased levels of competition influence athletes' decisions of engaging in unsanctioned instrumental aggression. Athletes continuously weigh between their perceived opportunity for success in the competition from engaging in unsanctioned aggression to obtain an edge versus probability and severity of punishment from unsanctioned aggression [13, 18, 19]. The team physician should at a minimum understand the rules of the sport at play, as well as the context and stakes of each match covered. It will be important to recognize if an athlete or team is on a losing streak and if there is a heightened need to win matches at the end of a regular season in order to advance to the next round or avoid relegation. An engaged team physician will be aware of situations where the opportunity for an edge from unsanctioned aggression may outweigh the punishment levied. This will allow the team physician to be prepared to respond to potential injuries stemming from these situations of increased competitive tension.

The role of social environment can be informed by social learning theory and length of time the athlete has participated in sports where on-field aggression is glorified and associated with success in competition [13, 16, 20]. Moral atmosphere is a well-researched determinant of aggression and violence in sports, specifically looking at behaviors of coaches, parents, and spectators [13, 21–23]. When these actors encourage aggressive behaviors during competition, research has shown an association with aggression among athletes; however, there is no clear indication on whether this is instrumental aggression or hostile/reactive aggression [13, 24]. Further, violence in sports doesn't always involve athletes directly. Referees are often at the center of conflict due to perceived poor officiating from one or both teams and may receive physical and/or verbal abuse from players, coaches, and spectators [13, 25–27]. Here again, knowledge of the context of each match covered is important. The team physician must be aware of team rivalries, recent match outcomes between the teams predisposing for increased tension, such as controversial officiating decisions, or recent history of previous altercations between teams or spectators. Knowledge of these factors may allow the team physician to adjust medical staff accordingly in preparation for these matches.

A recent high-profile incident in amateur sports occurred during the Women's 2024 Southeastern Conference Basketball Championship between Louisiana State University (LSU) and South Carolina, after a hostile/reactive aggression from a South Carolina player in the final two minutes of the game. Both benches subsequently cleared and stormed the court, and a spectator jumped over the score table onto the court as the brawl ensued. The spectator, brother of an LSU player, was rapidly removed by the police. The teams were separated, followed by multiple ejections and suspensions for players on both teams. The brother was charged with assault and battery as well as disorderly conduct. No injuries occurred; however, this incident garnered national attention as it occurred amid an increase in interest in viewership for women's college basketball. Invasion from spectators onto the playing field can happen even in televised and highly visible amateur sporting events where increased security personnel exist and consequences could be most severe. While rare, it is also possible for athletes to engage in physical altercations with fans, such as the "Malice at the Palace" incident when a professional basketball player of the Indiana Pacers went into the audience seats and threw punches at spectators. Knowledge that similar incidents may happen in amateur settings with less media visibility, fewer security measures in place, and where consequences could potentially be avoided will allow the team physician to approach these events with an appropriate level of vigilance and readiness. Understanding who local security is at events and how to work alongside our colleagues is essential.

Another high-profile incident, this time off the field and not involving athletes directly, occurred at a Spring 2024 7-on-7 youth football invitation-only camp and tournament in Atlanta, Georgia. This involved a former NFL quarterback and most valuable player (MVP) winner, who was the sponsor of one of the participating youth teams. This former athlete was involved in an altercation with two coaches of participating teams. Tournament staff and security intervened to separate the individuals, and no medical attention was necessary, nor were criminal charges pursued. However, it presents a prime example of the spectator role in youth sports and how incidents of violence divert the attention from youth sport development. Further, it emphasizes the moral atmosphere and the potential effect media portrayal can have on youth athletes to observe a high-profile role model involved in acts of aggression around youth sports, even if acting in self-defense. For the team physician, this serves as a reminder that any individual, regardless of profile, can be an aggressor and/or victim in a sporting event and that readiness is not only needed during an ongoing match but at all times during coverage of any sporting event.

It is also prudent to understand what type of injuries are most prevalent with increased violence. During striking sports, there are more injuries to the wrist and hands, while non-striking sports commonly involve trauma to the neck, shoulders, and elbows [28]. Adolescent violent injury in sports treated in the emergency department demonstrated most injuries were to the head and neck (52.2%) and almost a quarter of those injuries resulted in traumatic brain injuries, mostly from being shoved or hit. The extremities were also commonly injured (34%) with a large number of contusions, lacerations, and fractures. The playground was responsible for most injuries in children 9 years old or younger (65.2%), while bicycling (26.7%) and basketball (45.3%) were the culprits for 10- to 14-year-olds and for 15- to 19-year-olds, respectively. It was suggested to educate young athletes to prevent conflicts with peaceful resolutions, to practice safe sporting activity, and that parental supervision would be helpful [29].

The positive mental and physical health effects associated with participation in youth sports are well documented in the literature [10]. However, the destructive effects of interpersonal violence on athletes have been shown to counteract the protective factors on mental health that participation in sports confers [10, 30]. Athletes exposed to violence in sports present similar mental health issues to those who experience other forms of interpersonal violence outside of sports including anxiety, depression, PTSD, and self-injury [10, 30]. In the setting of youth sports, self-injury can present in the form of adopting unhealthy coping behaviors such as eating disorders, overtraining, or competing with injuries [3, 31]. These behaviors can subsequently lead to detrimental effects on performance and ultimately self-removal from participation in sports [3, 10].

Participation in youth sports has increased over the last few decades with an observed trend to early sports specialization [10, 32]. Early sports specialization has been associated with increased exposure to violence in sports as these athletes tend to train and compete during more hours and at a higher level of intensity and competition [10, 33]. That environment itself pushes the athlete to tolerate increased pressure from parents, coaches, and spectators in pursuit of finding success in competition. However, these athletes are not immune to the detrimental cumulative effects of exposure to violence in sports. While these athletes are pushed to achieve higher levels of performance, a concerted effort from coaches and parents is necessary to recognize that these athletes are exposed to more forms of aggression and violence in sports. Maintaining participation in sports and achieving higher levels of performance stem not only from surpassing physical performance barriers but also maintaining healthy behaviors with a focus on mental health. In 2016, the International Olympic Committee released a consensus statement addressing harassment and abuse in sport and emphasizing the need for a multiagency approach, including stakeholders such as sport organizations, athletes, sports medicine practitioners, and sports science researchers [3]. Sport organizations have a responsibility to educate parents, coaches, spectators on permissible behaviors during competition and clearly delineate and enforce consequences of violations [3]. Sports medicine practitioners must be trained in recognizing violence in sports and responding to it accordingly to establish trust with the athletes. Sports science researchers also play a role in growing the body of evidence surrounding detrimental effects of non-sanctioned aggression in sports and synthesizing information for widespread public awareness and implementation [3]. This role of the team physician may extend beyond the coverage of an event to help coordinate non-orthopedic or sports medicine care for athletes. For youth hockey athletes, one group of authors recommend the adoption of nonviolent role models and to educate the athletes on the consequences for aggression and the devastating effects of potential head trauma [34]. Updating rules to be safer and inflicting penalties on aggressors and teams has been shown to improve violent behavior. It has also been suggested to have medical experts involved for injury prevention and regular collaboration with all involved parties [34, 35].

9.5 Violence in Professional Sports

A recent high-profile example of violence in sports led to the longest game suspension in the history of the NFL for a single on-field incident [36]. On November 14, 2019, Week 10 of the regular season, the Cleveland Browns faced the Pittsburgh Steelers. In the final minute of the fourth quarter, which the Browns went on to win 21–7, the defensive end from the Browns sacked the quarterback of the Steelers after he completed his passing throw. They continued to grapple on the ground as the Steelers' quarterback unsuccessfully attempted to remove the opponent's helmet. The Browns' defensive end retaliated by successfully removing the Steelers' quarterback's helmet and subsequently struck him against the head with it. A Browns defensive tackle also received a one-game suspension due to his involvement in the incident, shoving the Steelers' quarterback to the ground [37]. The center from Pittsburgh received a three-game suspension for retaliating against the Browns' defensive end with punching and kicking [37]. The Steelers' quarterback did not receive a game suspension but was fined for his involvement. The suspended players went without pay and were fined; however, the Browns' defensive end received an indefinite game suspension which lasted the remaining six games of the regular season and postseason, for which the Browns did not qualify. In the NFL official statement, the sanctioning language included violating "unnecessary roughness and unsportsmanlike conduct rules, as well as fighting, removing the helmet of an opponent and using the helmet as a weapon" [37]. The league levied a hefty penalty for both franchises, with a $250,000 fine each [37]. Following the incident, no criminal charges were pressed. The Browns' defensive end was reinstated on February 12, 2020, ahead of the new season after his meeting with the NFL commissioner [38]. Various reports indicated the six-game suspension without pay amounted to approximately $1.14 million in lost salary [39]. The incident did not appear to have a lasting economic or career impact. In the summer of 2020, with 2 years left on his rookie contract, the Browns' defensive end went on to sign a 5-year extension with his team worth $125 million, for a deal totaling $144 million over 7 years [40]. This made him the highest-paid defensive player in annual average salary in NFL history and the highest paid NFL player in a non-quarterback position at the time [40].

While medical staff involvement was not necessary in the scenario above, it emphasizes how pieces of equipment necessary to participate in a specific sport can be unintentionally and, in this case, intentionally utilized to harm another athlete. A direct blow with a helmet to the unprotected head of an athlete could result in short- or long-term neurologic consequences. The team physician must be prepared in any amateur or professional setting to respond to these injuries which can be considered unnatural to the sport, but possible due to the inherent use of cleats' studs, bats, rackets, skate blades, and helmets among others. While beyond the scope of the team physician, it is important to consider how athletes may act in the future based on the amount of punishment and/or effects that actions of dangerous violence have on the player's career. Further, fans of professional sports often emulate their favorite superstars, which can affect the severity of violence in amateur sports [41].

A recent high-profile example of stadium violence with massive repercussions occurred on March 5, 2022, at the Estadio Corregidora in

Queretaro, Mexico, during a Liga MX Week 9 regular season soccer match between Queretaro Futbol Club and Atlas Futbol Club [42]. Atlas Futbol Club was up 1–0 on the scoreboard when fans from both teams invaded the pitch near the 60th minute of the match. Players and staff from both teams quickly ran to the locker rooms. This invasion stemmed from multiple brawls throughout the stadium between rivaling supporters. Media captured individuals launching trash cans and stadium seats and engaging in fist fights, with bloodied fans from the riot observed throughout the stadium. Rioters weren't the only ones invading the pitch. Groups of families were captured on television running onto the field, fleeing in panic from the violence. Following the brawl, the Municipal Civil Protection Department of Queretaro reported 22 injured, with nine individuals transported to a hospital, of which two were reported to be in critical condition [43]. No deaths were officially reported; however, controversy remains to date on the veracity of the official reports. The match was not renewed, and the Liga MX suspended all remaining Week 9 matches. This incident garnered international attention with the Confederation of North, Central America, and Caribbean Association Football (CONCACAF) and FIFA launching statements condemning the violence. Liga MX closed the Estadio Corregidora for an entire year forcing all Queretaro Futbol Club teams to play all home games without fans [42]. Queretaro's animation supporters or *barras* were banned from the stadium for three years and from away matches for one year. The team's ownership group was also fined $70,450 USD and banned from all Liga MX activities for five years [42]. Queretaro Futbol Club was returned to previous owners Grupo Caliente, who were also tasked by the Liga MX with selling the club within a set timeline. To date, Grupo Caliente still owns the club. Facial recognition technology and fan IDs were implemented for the following 2022–2023 Liga MX season across all stadiums following heavy criticism for local, municipal, and state authority response in arrests after the incident in Queretaro [44].

An important tenet of providing first aid to others is to assess the scene and ensure your own safety and the safety of your staff in order to appropriately provide care for others. In settings like these where chaos overwhelms the layers of security of an event, it is essential to exercise judgment to avoid harm while also assessing when and how to respond to the victims of the chaos. While ensuring the safety of your staff, it is essential to transmit calmness through preparedness and thus address the event systematically to avoid an unorganized response to chaos. Reminding your staff of these scenarios and reviewing appropriate levels of responses before covering an athletic event is of paramount importance. However, showcasing strong leadership and calmness through preparedness when a field invasion like the event described above occurs can ensure adherence from your staff to the protocols you have outlined and rehearsed.

9.6 Preparing for Potential Violence

When analyzing and preparing for potential violent injuries in a sporting event, we must take into account the degree to which violence may escalate in these events. Factors, including the geographic location, the number of people in attendance, the level of play, the tournament and tournament stage, and the history of long-standing rivalry between teams will affect how much violence is likely to escalate [45].

Therefore, if talking about a youth league or a high school event, the potential threats and level of violence are likely much diminished compared to that of a championship game between two rival teams at a high-stakes tournament.

In a much lower-stakes situation like those in youth leagues, one should consider physical altercations between players and player injuries and minor altercations between the audience members. In bigger events, we not only take those factors into account, but we must add the potential of violent injury during entry and exit from the sporting venue at the gates, outside of the venue, and potential terrorist attacks.

In both scenarios, the roles of governing bodies within a sport and at a legislative level show efficacy in the prevention of violence [46, 47]. Regulations at a legislative level have been suggested to set forth safety standards for stadiums and facilities and mandate that sports governing bodies have self-regulating policies in place [46]. Different associations are able to set rules and fines that are not only costly to players who commit violent acts but are also punitive to fans for aggressive behavior at the stadiums. Furthermore, governing bodies are in charge of determining the equipment that is permissible to players to avoid further injury or prevent injury, and they also set forth safety protocols for different scenarios within the venue to protect players and fans alike.

9.7 Regulations

Among the most common types of regulations in all sporting events is the distribution of penalties, fines, and suspension for aggressive behavior most commonly in the form of fouls. The penalty needs to be commensurate to the behavior or to the potential encouragement for aggressive behavior [45]. There are certain scenarios within sporting events that encourage violent behavior, take as an example a soccer match in the 90th minute, with both teams tied, and an almost certain scoring opportunity for a striker. A defender in this scenario may prefer resorting to violence and committing a tactical foul to prevent a loss and go into extra time or a penalty shootout. If the punishment for this was a mere yellow card for the players committing the foul, it would possibly be worth the potential of injuring the player, but a red card, in which they are kicked out of the game and miss the next game, is more likely to discourage the defender from committing the foul. Thus, preventing violence on the pitch.

To prevent violence between players off the pitch, similar practices can commonly be employed. The use of cards to have players miss future games should and is often employed retroactively through video review in settings where violence erupted after a game ended. In the setting of professional sports, monetary fines to both the player and the club can also be added to discourage players and coaching team members from engaging in violent behavior.

Controlling violence outside of pitch and in the stands is a much different challenge but just as important. In order to better control fans and prevent aggressive behavior, governing bodies can employ numerous regulations and rules.

Rules can be set on the stadium in order to make them eligible to host games. These rules may be as simple as limiting the sale of alcohol or using plastic cups over the use of glass bottles [45]. However, other important measures to ensure safety include enforcing the use of metal detectors and bag screening prior to entering the venue, employing an adequate number of security guards posted strategically along entrances to different seating sections, ensuring a perimeter large enough around the venue to prevent people without tickets from entering the venue, having proper identification cards, reducing the patrol perimeters for staff so they don't have to cover as much area, reducing bleacher space and replacing it for assigned seating, and delineating fan-specific areas of the stadium with buffer zones in between different fan areas [47]. The team physician should be aware of these posts and how to navigate throughout the stadium.

Once in the stadium venues may apply a similar penalty system for fan behavior on the teams. For example, if fans throw objects at players of the opposing team during the middle of the game, the team they root for may get a monetary fine. This would essentially incite the team to encourage fans to be on their best behavior during matches. Furthermore, certain protocols need to be in place in case of unruly behavior. These may range from removing inciting individuals from the stadium and banning them to canceling and postponing matches due to behavior or threats that may pose a risk to the safety of spectators. Delays may encourage crowds to settle down, with the idea that they will be rewarded by the continuation of the match.

Violence is often prevented before the consideration of gameplay. Most of the literature supports education, updated rules, and policies favoring nonviolence. This can come from an individual team culture as well as the governing body organization of the sport. Coaches should be encouraged to learn about violence in sports and conflict management strategies to help create an environment for athletes to thrive [12, 44]. Athletes should also learn about appropriate conflict management strategies. This could be through the team, community outreach programs, and/or ideally a class in primary schools. Training programs against violence for athletes, coaches, referees, and ancillary staff can help limit non-accidental violent sports behaviors. Communication between all involved parties is vital, and roles should be clearly delineated [10]. Safeguard systems should be in place for anyone to safely and anonymously report misconduct as well as psychological, physical, and sexual violence. For youth athletes, the team physician, as a mandatory reporter, should take notice of any signs of child abuse and report this to the appropriate channels without reservation. Prevention of violence is always preferred over creating policies after the damage has already been done. Awareness campaigns at all levels of athletics can help promote sportsmanship and respect.

9.8 Safety Equipment

Another way for safety to be implemented in sports settings is through the equipment worn by players. Depending on the sport, mandatory protective gear such as helmets, shoulder pads, mouth guards, knee pads, etc. needs to be made mandatory. This is especially important in contact sports that are violent in nature, such as American football or rugby. While the use of helmets is mandatory in American football, but the use of headgear is only optional in rugby, there is a higher incidence in head injuries in rugby, which has been attributed to the difference in rules [48]. The rules of many sports should and often do allow for additional protective equipment for players that may have returned to play but are still healing from an injury. These may include masks to protect players with a healing nasal fracture, allowing for casts and splints in certain scenarios or allowing for helmets in a sport where those aren't typically worn. Such is the example of, a goalkeeper for the English premier league team Chelsea, who sustained a depressed skull fracture in a match against Reading when a midfielder's knee collided against his head in the early minutes of the match [49]. After his recovery, the Chelsea goalkeeper returned to play with a protective head guard, which he wore for the duration of his career. The infrastructure around the players is also important. In stadium settings, the placement of fences, as well as a roof over the substitute benches, also help protect players from violence coming from the spectators. Furthermore, an enclosed locker room space that is not easily accessible for the fans is not only helpful for the safety of the players but is helpful for a team physician to be able to evaluate and treat a patient in a covered and enclosed area. If a dislocation/fracture reduction or orthopedic maneuver needs to be performed, the team physician should make his or her best effort to perform this out of view from players, fans, and the media.

9.9 Technological Advances

Over time advances in technology have made injury prevention in sports better. This has been in the form of technological improvements in protective gear, wearable technology for data acquisition and motion tracking, as well as technological advances in refereeing. When looking at the design of headgear over the last century, there is a clear progression toward safety-focused modifications. Take, for example, football helmets. In the 1920s, helmets were mere leather coverings, but as time progressed, helmets became hard shelled, incorporated a face shield, and even became multilayered to better absorb impact [48]. The technological advances have also focused on the sport itself. In the case of helmets, the design of a football helmet is much different than that of a hockey helmet. In the case of the football helmet, the design is focused on reducing the impact of lateral blows and the helmet is larger and covers the ears; in hockey, the helmet has no ear coverings, it is smaller, and the face shield protects the eyes and is transparent [48]. Other protective equipment has focused on becoming lighter to allow for better sports performance while maintaining or bettering safety parameters for the player. Team physicians need to stay up to date on the latest technology that is supported to help prevent injuries to their athletes.

Wearable technology has also seen an increased use with the purpose of injury prevention [50]. Wearable technology can provide helpful data in order to prevent an athlete from engaging in injury-prone behavior. This can also help develop training programs to improve their technique to reduce chances of injury [51]. Wearable technology does this by tracking movements and identifying poor motor coordination that can later be worked on. It also helps estimate when athletes are reaching high training loads, which may make them more prone to injury. Wearable technology for motion tracking has been used in baseball in which a wearable sleeve effectively tracks shoulder motion, and the data is used for injury prevention as well as for the design of targeted training plans for athletes [50, 51]. Team physicians have the opportunity to collaborate with sports scientists when analyzing this data.

Technological advances in refereeing have also allowed for better rule enforcement, which is not only important for the outcome of the sporting event but is also important in reducing injury through behavior. The use of video review and tracking technology has had the benefit of punishing aggressive behavior either through a card or through a fine that would have otherwise been missed, thus making players more conscientious of their acts on and off the pitch [52].

The use of video footage has also enhanced safety within the stadium. Widespread camera placement in a stadium paired with a closed-circuit monitoring system allows stadiums to have better surveillance to identify areas of conflict build up, as well as face tracking to identify persons of interest, run databases, and send pictures to the authorities if needed [47].

9.10 Terrorism Risks and Responses in Sports

Large gatherings of people are subject to risk in multiple forms, including environmental, individual, public health, psychological, and management [53]. Management includes planning proper safety precautions to prevent acts of terrorism and having a plan for emergency response in the case that one occurs. A study in 2023 analyzed the global terrorism database from 1970 to 2019 and identified 74 acts of terrorism at sporting events, with the majority at soccer events (44.6%), involving bombing/explosion (87.8%), and in the Middle East or North Africa [54].

One of the most infamous recent mass causality incidents in the United States was during the 2013 Boston Marathon. Almost 27,000 athletes were registered to compete in this historic race with over half a million spectators cheering along the 26.2-mile course. Tragically, this terrorist attack occurred in the final meters next to the finish line. Several hours into the race, two sequential bombs exploded in this highly populated finish-line area. Three spectators died instantly, and 281 people were injured with varying severity of trauma, including amputations and shrapnel injuries [55, 56].

As a team or covering physician of this event, it is time to take immediate action. It is vital to assess the scene for safety. Will there be another attack? Are there collapsed structures, shrapnel, or debris impeding your ability to get to the injured patients? How can you help smoothly transport patients for required escalated care? Is the local healthcare system aware of the event? The city and healthcare system of Boston performed a remarkable crisis response that led to excellent outcomes. The terrorist attack occurred between Boston's five adult level 1 trauma centers and four pediatric trauma centers [55]. Parallel to military medicine's structure, the emergency medical services (EMS) were the first to respond and helped place an in-field tourniquet on 83% of patients with hemorrhaging extremity injuries. It is possible that many of these may have been improvised given the number of patients needed to be treated, in which improvisation is a necessary principle a team physician must possess. The team physician must also be cognizant of potential infectious transmission with multiple bloody patients and should use personal protective equipment (PPE). Ambulances helped evacuate the injured on the scene along with the police as well as other vehicles. Per the Boston EMS central command structure, patients with an Injury Severity Score of 16 or higher were relatively spread out over the multiple trauma centers within a median time of 11 min. These centers were prepared for a disaster scenario such as this. Their respective Hospital Incident Command System centers helped transfer patients out of the emergency departments in order to make room for the influx of injured patients. No other elective surgical case was started after notification of the attack [55].

Blast injuries from bombings can also pose additional problems with immediate care. Blast injuries can be classified as primary, secondary, tertiary, or quaternary based on how damage was conferred to the victim [57]. Primary injuries are due to the wave damaging the victim and impact hollow viscus organs, such as the tympanic membrane, gastrointestinal organs, and lungs [57]. Damaged tympanic membranes present the additional challenge of limiting the victim from hearing where aid may be coming from and where future danger is presenting. Damaged lungs can limit victim's endurance from escape. Needle decompression for tension pneumothorax and/or intubation may be necessary, which the team physician should be prepared to perform. Secondary blast injuries are penetrating trauma by projectiles, and important early management can revolve around stopping bleeding and maintaining circulation. Tertiary blast injuries occur when a victim is moved by the blast wave damaged by their surroundings, such as being knocked into a wall. Finally, quaternary injuries describe all others, such as burns, amputations, and fractures of the bones [57]. Secondary through quaternary blast injuries were prevalent in the example of the Boston Marathon bombing, and initial stabilization of fractures is key in early patient management to prevent further soft tissue and neurovascular injury.

127 of the 281 total treated patients were provided care at Level 1 trauma centers on the day of the bombing. The emergency departments were able to discharge 52 (41%) of the patients, while 75 (59%) were admitted, with 45 of the admitted patients (60%) being directly sent to the operating room for urgent surgery. The surgeries were mainly performed by the orthopedic and general trauma surgery teams. Forty-seven patients needed multiple surgeries, and 12 patients required a lower extremity amputation. No additional fatalities occurred [55] (Table 9.1).

Gates et al. recommend that all resident and fellow education have formal training for disaster operations [55]. The authors of this textbook share this ideology and developed this textbook to prepare team physicians for any environmental condition. An additional recommendation that Gates et al. described was to have healthcare organizations intimately involved with larger security forces, such as SWAT teams [55]. One of the authors of this chapter, Dr. Alfonso Mejia, works with a local SWAT team as a Tactical Emergency Response physician, team member, and medical director for the team. During highly visible events such as the Chicago Marathon, his team partners with other teams to provide tactical support throughout the course of the race (Figs. 9.2 and 9.3). This unique integration of police and medical operations offers seamless medical care coordination. SWAT teams regularly practice and stage potential disaster scenarios to practice the initial disaster response. Repeatedly reviewing and practicing your team physician's medical protocols for dangerous situations is prudent for overcoming possible roadblocks and performing best quality patient care. Having a liaison with law enforcement is also important, especially during large and international events. Long-term recovery must also be considered to provide support for the victims as well as to rebuild confidence to the public.

Table 9.1 Operative room interventions and hospital course for injured patients at the Boston Marathon bombing

OR interventions, inpatient phase, and discharge disposition	
	No. (%) Patients
Patients who went directly to OR, April 15	45
Patients who went to ICU, then OR, April 15	6
Patients who went to ward, then OR, April 15	3
No. procedures per patient, April 15	15
1–2	15
3–5	29
6–10	9
>10	1
Patients who underwent operative procedures, April 15	
Craniotomy	0
Thoracotomy	0
Laparotomy	4
Foreign body removal	24
External fixator	7
Soft tissue debridement	54
Vascular procedure	14
Lower extremity amputation	12
Fracture repair	15
Patients who underwent repeat operative procedures	47
Patients who underwent amputation after April 15	3
ICU LOS, median	4 d
ICU LOS, range	1–22 d
Patients who experienced complications	7 (5.5)
Hospital LOS, median	8.5 d
Hospital LOS, range	1–49 d
Discharge location	
Rehabilitation	33 (44.0)
Home	30 (40.0)
Home with services	11 (14.7)
Skilled nursing facility	1 (1.3)

Reproduced from Gates et al., *The Initial Response to the Boston Marathon Bombing: Lessons Learned to Prepare for the Next Disaster*, Ann Surg., 260 (6): 960–966, 2014, doi:10.1097/SLA.0000000000000914, with permission

Fig. 9.2 Large group of runners at the Chicago marathon

Fig. 9.3 SWAT team preparing for the Chicago marathon

9.11 Preventive Measures Against Terrorism

Initial counterterrorism measures revolve around surveillance and detection [58]. However, in the event that threats are not prevented, security staff should be adequately trained and positioned to respond to a threat and evacuate the public while helping those who may be injured.

Preventative measures against terrorism begin with limiting danger that can be brought to the event and developing a plan to prepare for the unexpected. The venue of the event can have a large impact on what sort of security can be implemented. For example, a closed-off high school gymnasium will have fewer entrances, a smaller crowd, and a more contained space to observe than an outdoor professional golf course or countrywide bike race [59]. Team physicians should be sure to consider the specific venue for the event they are covering and how that will impact response to disaster. Initial steps for security include a bag check, a metal detector, and surveillance for the venue. More advanced technology, such as explosive trace detection, can be implemented as needed for higher-risk events [60]. After screening, the next steps in counterterrorism involve implementing a collaborative plan between public health, law enforcement, and healthcare providers to communicate to the public, to evacuate, to provide healthcare, and to limit additional threats.

A crucial aspect of terrorism planning and response is disseminating information to the public and having a protocol in place for their response. Two theories can be utilized to explain human response to threats: the protective motivational theory (PMT) and the elaboration model likelihood of persuasion (ELM). PMT states that people will respond to a threat based on the level of perceived danger and their ability to avoid, or cope with, the danger [61]. ELM describes two cognitive routes that influence decision-making: a central route based on objective information and a peripheral route based on cues and perceptions [62]. Communication systems designed around these explanations of human behavior can maximize human response to threats and optimize a terrorism response protocol. For example, the central route is a more predictable response, and therefore, information should be presented with objective information that accurately describes what threat is present and how the public can avoid said threat. By providing information both on the perceived danger and how to cope, civilians are best equipped to respond as instructed by law enforcement. Current research on this topic is ongoing to further optimize how to maximize response from the public and avoid notification fatigue [63]. Team physicians can benefit from understanding how people respond to threats to optimize the way they provide direction and caution in situations when a threat is present and should be familiar with PMT and ELM theories.

Along with communication, the initial response to terrorism necessitates multiple teams working together to reduce the risk of further threat, provide care for injured individuals, and safely evacuate the public from the scene. Teamwork between law enforcement and healthcare providers is crucial and can be optimized through training. While training is important, it is not ubiquitous among healthcare provider training programs. A study on the incorporation of terrorism preparedness training in emergency medicine, family practice, and pediatric residency programs identified not all programs incorporate training, and less than 50% of programs specifically describe emergency responses for pediatric victims [64]. While the implementation of training programs has not yet become universal, the efficacy of these programs has been demonstrated to significantly improve provider preparedness to act in the case of emergency [65]. Organizations such as the Joint Counterterrorism Awareness Workshop Series, sponsored by the Department of Homeland Security, host workshops to help coordination between the public and private sectors including healthcare and law enforcement [66]. We recommend team physicians seek out a workshop in counterterrorism to prepare themselves and develop a protocol to respond to potential disaster.

A key factor in emergency preparedness is the teamwork between healthcare providers and law enforcement. The Big Five theory describes team leadership, mutual performance monitoring, backup behavior, adaptability, and team orientation as the cornerstones for effective teamwork during stressful situations [67]. Training situations can enhance these skills within the team by developing mutual trust and facilitating practice with effective communication styles such as closed-loop communication, which is crucial for effective response to terrorism [68]. In addition to supporting each other to provide optimal aid to victims, collaboration between responders is also important for enhancing their resilience after being exposed to the devastation of terrorism [69]. Team physicians can enhance their preparedness for terrorism disasters by familiarizing themselves with local law enforcement and public health contacts. These skills can translate to everyday jobs of the team physician to improve communication and outcomes.

Training should equip all responders with a plan to execute in the case of an emergency. Effective emergency action plans necessitate all team members understanding their role within the team and an identifiable incident commander directing the team [59]. Often, the initial response involves a combination of evacuation and medical stabilization of acute injuries. However, the treatment of injuries after a terrorist incident can be limited by the need to evacuate. Therefore, when compared with victims of other mass casualties, victims of terrorism events can present to nearby hospitals less medically stable and require more care in the emergency department [70].

Historically, the preferred response to terrorism is "stage and wait," which involves not rushing into a scene after an explosion due to the potential danger of a second. Recently, this approach has been called into question due to the relative infrequency of multiple bombings. An ethical question with no objective answer is therefore raised: is the potential benefit from immediate aid to the public outweighed by the rare but significant potential for damage to first responders when rushing the scene? A study reviewed the global terrorism database and identified that 70 per 1000 incidents involved a second bomb [71]. Situational considerations should impact emergency personnel response to these situations, such as the magnitude of the explosion, individuals in peril, and safety of routes to provide aid for emergency personnel on staff.

While certain factors, such as crowd size, can obscure threat detection, advanced technologies can be utilized to identify risks before danger occurs [72]. Technology utilized for terrorism has also developed through the years, and in addition to explosives, plans should be implemented for alternative threats such as chemical, biological, radiological, and nuclear [72]. Machine learning has been proposed to help prevent terrorism by utilizing data to predict where threats may occur and allowing personnel to adjust security to prevent those threats from coming to fruition [73]. Further implementation of explosive trace detection, artificial intelligence, and machine learning will best equip public health to predict and avoid terrorist events at mass gatherings.

9.12 Conclusion

Whether it is a "Malice at the Palace" or a verbal altercation at a Little League Baseball Game between opposing parents and a child on your team, the prepared team physician will know how to intervene and/or who to notify to prevent future issues. Understanding and practicing emergency protocols is required to be ready for any potential danger. Carrying an updated team physician game bag with medications, splints, laceration repair kits, airway management supplies, PPE, etc. should be standard of care when covering events. Future research should look at implementing specific protocols or education that are supported by the literature in reducing violence. It is important for continued vigilance and improvement in addressing danger and violence in sports.

9.13 Key Takeaways

- Violence in sports is a multifaceted issue that occurs at all levels of athletics, in all age groups. These experiences can lead to lifelong comorbidities, such as depression, anxiety, post-traumatic stress disorder, eating disorders, and more.
- The team physician must be aware of the personal, cultural, and environmental conditions that may encourage violent conditions. This may include the limitations of physical contact within the rules of a sport, historical violent events between players and/or teams, the current culture of the team, and external pressures on an athlete (e.g., coaches, parents, fans). Knowledge of potential triggers and likely types of injuries can allow the team physician to remain vigilant and provide quick and safe care.
- Best practices involve evaluating and treating an injured athlete in a protective and enclosed area. If a dislocation/fracture reduction or orthopedic maneuver needs to be performed, team physicians should make their best effort to perform this out of view from players, fans, and the media.
- When covering international or high-profile events, the team physician must coordinate with security, the event coordinators, and tertiary medical centers in preparation for a dangerous scenario.

References

1. Field M, Collins MW, Lovell MR, Maroon J. Does age play a role in recovery from sports-related concussion? A comparison of high school and collegiate athletes. J Pediatr. 2003;142(5):546–53.
2. Medical planning timeline for mass sporting event—UpToDate [Internet]. [cited 2024 Sep 23]. Available from: https://www.uptodate.com/contents/image?imageKey=SM%2F109988&topicKey=SM%2F13816&source=see_link
3. Mountjoy M, Brackenridge C, Arrington M, Blauwet C, Carska-Sheppard A, Fasting K, et al. International Olympic Committee consensus statement: harassment and abuse (non-accidental violence) in sport. Br J Sports Med. 2016;50(17):1019–29.
4. Gattis C, Moore M. A conceptual analysis of maltreatment in sports: A sport social work perspective. Front Sports Act Living. 2022;4:1017308.
5. Kerr G, Battaglia A, Stirling A. Maltreatment in Youth Sport: A Systemic Issue. 2019 Aug 1 [cited 2024 Aug 30]; Available from: https://journals.humankinetics.com/view/journals/krj/8/3/article-p237.xml
6. Logan K, Cuff S, Council on Sports Medicine and Fitness, LaBella CR, Brooks MA, Canty G, et al. Organized Sports for children, preadolescents, and adolescents. Pediatrics. 2019;143(6):e20190997.
7. Taliaferro LA, Rienzo BA, Miller MD, Pigg RM, Dodd VJ. High school youth and suicide risk: exploring protection afforded through physical activity and sport participation. J Sch Health. 2008;78(10):545–53.
8. Merrick MT. Vital Signs: Estimated Proportion of Adult Health Problems Attributable to Adverse Childhood Experiences and Implications for Prevention—25 States, 2015–2017. MMWR Morb Mortal Wkly Rep [Internet]. 2019 [cited 2024 Aug 19]; 68. Available from: https://www.cdc.gov/mmwr/volumes/68/wr/mm6844e1.htm
9. Bellis MA, Hughes K, Ford K, Rodriguez GR, Sethi D, Passmore J. Life course health consequences and associated annual costs of adverse childhood experiences across Europe and North America: a systematic review and meta-analysis. Lancet Public Health. 2019;4(10):e517–28.
10. Daignault I, Deslauriers-Varin N, Parent S. Profiles of teenage athletes' exposure to violence in sport: an Analysis of their sport practice, athletic behaviors, and mental health. J Interpers Violence. 2023;38(11–12):7754–79.
11. Dallam SJ, Ortiz AJ, Timon CE, Kang JS, Hamilton MA. Interpersonal violence in elite U.S. athletes: Prevalence and mental health correlates. Journal of Aggression, Maltreatment & Trauma. 10:1–19.
12. Vertommen T, Kampen J, Schipper-van Veldhoven N, Wouters K, Uzieblo K, Van Den Eede F. Profiling perpetrators of interpersonal violence against children in sport based on a victim survey. Child Abuse Negl. 2017;63:172–82.
13. Spaaij R, Schaillée H. Unsanctioned aggression and violence in amateur sport: A multidisciplinary synthesis. Aggress Violent Behav. 2019;44:36–46.
14. Danioni F, Barni D. The relations between adolescents' personal values and prosocial and antisocial behaviours in team sports. Int J Sport Exerc Psychol. 2019;17(5):459–76.
15. Conroy DE, Silva JM, Newcomer RR, Walker BW, Johnson MS. Personal and participatory socializers of the perceived legitimacy of aggressive behavior in sport. Aggress Behav. 2001;27(6):405–18.
16. Huang DB, Cherek DR, Lane SD. Laboratory measurement of aggression in high school age athletes: provocation in a nonsporting context. Psychol Rep. 1999;85(3 Pt 2):1251–62.
17. Moesch K, Birrer D, Seiler R. Differences between violent and non-violent adolescents in terms of sport

background and sport-related psychological variables. Eur J Sport Sci. 2010;10:319–28.
18. Abrams M. Anger management in sport: understanding and controlling violence in athletes, vol. xiii. Champaign: Human Kinetics; 2010. p. 282. (Anger management in sport: Understanding and controlling violence in athletes)
19. Sherrill A, Bradel L. Contact sport participation predicts instrumental aggression, not hostile aggression, within competition: quasi-experimental evidence. J Aggress Maltreat Trauma. 2017;9:50–7.
20. Kreager D. Unnecessary roughness? School sports, peer networks, and male adolescent violence. Am Sociol Rev. 2007;72:705–24.
21. Shields DL, LaVoi NM, Bredemeier BL, Power FC. Predictors of poor sportspersonship in youth sports: Personal attitudes and social influences. 2007 Dec 1 [cited 2024 Aug 27]; Available from: https://journals.humankinetics.com/view/journals/jsep/29/6/article-p747.xml
22. Bortoli L, Messina G, Zorba M, Robazza C. Contextual and individual influences on antisocial behaviour and psychobiosocial states of youth soccer players. Psychol Sport Exerc. 2012;13(4):397–406.
23. Walters S, Schluter P, Stamp D, Thomson R, Payne D. Coaches and referees' perspectives of sideline behaviour in children's team Sports: A cross-sectional study. IJSMaRT. 2016;23:51–74.
24. Nicholson M, Hoye R. Contextual factors associated with poor sport spectator behaviour. Manag Leis. 2005;10(2):94–105.
25. Shapcott K, Bloom G, Loughead T. An initial exploration of the factors influencing aggressive and assertive intentions of women ice hockey players. Int J Sport Psychol. 2007;38:145–62.
26. Traclet A, Romand P, Moret O, Kavussanu M. Antisocial behavior in soccer: A qualitative study of moral disengagement. Int J Sport Exerc Psychol. 2011;9(2):143–55.
27. Ackery AD, Tator CH, Snider C. Violence in Canadian amateur hockey: the experience of referees in Ontario. Clin J Sport Med. 2012;22(2):86–90.
28. Noh JW, Park BS, Kim MY, Lee LK, Yang SM, Lee WD, et al. Analysis of combat sports players' injuries according to playing style for sports physiotherapy research. J Phys Ther Sci. 2015;27(8):2425–30.
29. Conn JM, Annest JL, Bossarte RM, Gilchrist J. Nonfatal sports and recreational violent injuries among children and teenagers, United States, 2001–2003. J Sci Med Sport. 2006;9(6):479–89.
30. Finkelhor D, Ormrod RK, Turner HA. Poly-victimization: a neglected component in child victimization. Child Abuse Negl. 2007;31(1):7–26.
31. Boudreault V, Gagnon-Girouard MP, Carbonneau N, Labossière S, Bégin C, Parent S. Extreme weight control behaviors among adolescent athletes: links with weight-related maltreatment from parents and coaches and sport ethic norms. Int Rev Sociol Sport. 2022;57(3):421–39.
32. Jayanthi NA, Post EG, Laury TC, Fabricant PD. Health consequences of youth sport specialization. J Athl Train. 2019;54(10):1040–9.
33. Vertommen T, Kampen J, Schipper-van Veldhoven N, Uzieblo K, Van Den Eede F. Severe interpersonal violence against children in sport: associated mental health problems and quality of life in adulthood. Child Abuse Negl. 2018;76:459–68.
34. Cusimano MD, Ilie G, Mullen SJ, Pauley CR, Stulberg JR, Topolovec-Vranic J, et al. Aggression, violence and injury in minor league ice hockey: avenues for prevention of injury. PLoS One. 2016;11(6):e0156683.
35. Cusimano MD, Nastis S, Zuccaro L. Effectiveness of interventions to reduce aggression and injuries among ice hockey players: a systematic review. CMAJ. 2013;185(1):E57–69.
36. ESPN.com [Internet]. 2019 [cited 2024 Aug 30]. Browns' Garrett gets indefinite ban from NFL. Available from: https://www.espn.com/nfl/story/_/id/28087446/browns-myles-garrett-suspended-indefinitely-steelers-maurkice-pouncey-gets-3-game-ban
37. NFL.com [Internet]. [cited 2024 Aug 30]. Browns DE Myles Garrett suspended indefinitely. Available from: https://www.nfl.com/news/browns-de-myles-garrett-suspended-indefinitely-0ap3000001076995
38. NFL.com [Internet]. [cited 2024 Aug 30]. Myles Garrett reinstated after indefinite suspension. Available from: https://www.nfl.com/news/myles-garrett-reinstated-after-indefinite-suspension-0ap3000001101817
39. NBC Sports [Internet]. 2019 [cited 2024 Aug 30]. Myles Garrett will lose at least $1.139 million. Available from: https://www.nbcsports.com/nfl/profootballtalk/rumor-mill/news/myles-garrett-will-lose-at-least-1-139-million
40. ESPN.com [Internet]. 2020 [cited 2024 Aug 30]. Source: Browns, DE Garrett finalizing $125M deal. Available from: https://www.espn.com/nfl/story/_/id/29461044/sources-browns-closing-5-year-125m-deal-myles-garrett
41. Lev A, Tenenbaum G, Eldadi O, Broitman T, Friedland J, Sharabany M, et al. "In your face": the transition from physical to symbolic violence among NBA players. PLoS One. 2022;17(5):e0266875.
42. ESPN.com [Internet]. 2022 [cited 2024 Aug 30]. Liga MX bans Queretaro fans, "barras" after brawl. Available from: https://www.espn.com/soccer/story/_/id/37626148/liga-mx-queretaro-home-fans-banned-1-year-barras-get-3-year-ban-owners-sell-club
43. ESPNdeportes.com [Internet]. 2022 [cited 2024 Aug 30]. Reporta Protección Civil del Estado de Querétaro 22 personas lesionadas, tras trifulca en Gallos vs Atlas. Available from: https://espndeportes.espn.

com/futbol/mexico/nota/_/id/10026376/queretaro-vs-atlas-liga-mx-pelea-heridos-sin-reporte-oficial-muertos

44. Guerra HRÁ. Grupo Milenio. 2023 [cited 2024 Aug 30]. Se cumple un año de la tragedia en La Corregidora de Querétaro. Available from: https://www.milenio.com/deportes/liga-mx/se-cumple-un-ano-de-incidente-violento-en-la-corregidora-en-queretaro
45. Tenenbaum G, Stewart E, Singer RN, Duda J. Aggression and violence in sport: an ISSP position stand. J Sport Psychol 1997; 11(1):1–7, 1.
46. Hicks WM. Preventing and punishing player-to-player violence in professional Sports: the court system versus league self-regulation*. J Leg Aspect Sport. 2001;11(2):209–24.
47. Hall S A. An examination of British sport security strategies, legislation, and risk management practices. The Sport Journal [Internet] 2010 Apr 9 [cited 2024 Sep 2]; 24(hin). Available from: https://thesportjournal.org/article/an-examination-of-british-sport-security-strategies-legislation-and-risk-management-practices/
48. Bonfield CM, Shin SS, Kanter AS. Helmets, head injury and concussion in sport. Phys Sportsmed. 2015;43(3):236–46.
49. Reading 0–1 Chelsea | Football | The Guardian [Internet]. 2020 [cited 2024 Sep 2]. Available from: https://web.archive.org/web/20200301104303/https://www.theguardian.com/football/2006/oct/15/match.chelsea
50. Seçkin AÇ, Ateş B, Seçkin M. Review on Wearable Technology in Sports: concepts, challenges and opportunities. Appl Sci. 2023;13(18):10399.
51. Zadeh A, Taylor D, Bertsos M, Tillman T, Nosoudi N, Bruce S. Predicting Sports Injuries with Wearable Technology and Data Analysis. Inf Syst Front [Internet]. 2021 Aug [cited 2024 Sep 2]; 23(4):1023–37. Available from: https://link.springer.com/10.1007/s10796-020-10018-3
52. Ferreira Meneguite YN, Bernardes Leite L, Correia Da Silva D, Gomes De Moura A, Neiva Lavorato V. Influence of the video assistant referee (VAR) on the Brazilian Men's soccer championship. JPES. 2022;22(4)
53. Tavan A, Tafti AD, Nekoie-Moghadam M, Ehrampoush M, Vafaei Nasab MR, Tavangar H, et al. Risks threatening the health of people participating in mass gatherings: A systematic review. J Educ Health Promot. 2019;8:209.
54. Rahman GR, Jasani GN, Liang SY. Terrorist attacks against sports venues: emerging trends and characteristics spanning 50 years. Prehosp Disaster Med. 2023;38(3):366–70.
55. Gates JD, Arabian S, Biddinger P, Blansfield J, Burke P, Chung S, et al. The initial response to the boston marathon bombing. Ann Surg. 2014;260(6):960–6.
56. Hanfling D. Boston bombings and resilience—what do we mean by this? Front Public Health. 2014;2:25.
57. Ferreri TG, Weir AJ. EMS Improvised Explosive Devices and Terrorist Activity. In: StatPearls [Internet]. Treasure Island (FL): StatPearls Publishing; 2024 [cited 2024 Sep 2]. Available from: http://www.ncbi.nlm.nih.gov/books/NBK482191/
58. Infections I of M (US) F on E, Knobler SL, Mahmoud AA, Pray LA. Scientific and Policy Tools for Countering Bioterrorism. In: Biological Threats and Terrorism: Assessing The Science and Response Capabilities: Workshop Summary [Internet]. National Academies Press (US); 2002 [cited 2024 Sep 2]. Available from: https://www.ncbi.nlm.nih.gov/books/NBK98397/
59. Rubin AL. Safety, security, and preparing for disaster at sporting events. Curr Sports Med Rep. 2004;3(3):141–5.
60. To KC, Ben-Jaber S, Parkin IP. Recent developments in the Field of explosive trace detection. ACS Nano. 2020;14(9):10804–33.
61. Floyd DL, Prentice-Dunn S, Rogers RW. A meta-analysis of research on protection motivation theory. J Applied Social Pyschol. 2000;30(2):407–29.
62. Petty RE, Briñol P. Emotion and persuasion: cognitive and meta-cognitive processes impact attitudes. Cogn Emot. 2015;29(1):1–26.
63. Cadet J, Sutherlin S, Burke RV. Public safety messaging during super bowl LVI: A pilot study and modified framework. PLoS One. 2023;18(4):e0284921.
64. Martin SD, Bush AC, Lynch JA. A national survey of terrorism preparedness training among pediatric, family practice, and emergency medicine programs. Pediatrics. 2006;118(3):e620–6.
65. Gershon RRM, Gemson DH, Qureshi K, McCollum MC. Terrorism preparedness training for occupational health professionals. J Occup Environ Med. 2004;46(12):1204–9.
66. Shapiro GL, Armstrong JH, Roberts K, Gordon J, Smith ER, Afman J, et al. The joint counterterrorism awareness workshop series (JCTAWS): integrating disciplines for enhanced capabilities during a complex coordinated attack. J Emerg Manag. 2019;17(3):210–2.
67. Espevik R, Johnsen BH, Saus ER, Sanden S, Olsen OK. Teamwork on patrol: investigating teamwork processes and underlaying coordinating mechanisms in a police training program. Front Psychol. 2021;12:702347.
68. Butler JC, Cohen ML, Friedman CR, Scripp RM, Watz CG. Collaboration between public health and law enforcement: new paradigms and partnerships for bioterrorism planning and response. Emerg Infect Dis. 2002;8(10):1152–6.
69. Paton D, Violanti JM. Law enforcement response to terrorism: the role of the resilient police organization. Int J Emerg Ment Health. 2008;10(2):125–35.

70. Pfenninger EG, Klingler W, Keiloweit T, Eble M, Wenzel V, Krüger WA. Terrorist attack training exercise-what can be learned?: Baden-Württemberg counterterrorism exercise (BWTEX). Anaesthesist. 2020;69(7):477–86.
71. Ragoler M, Radomislensky I, Dolev E, Renert L, Peleg K. Rethinking the current "stage-and-wait" paradigm. Prehosp Disaster Med. 2023;38(2):185–92.
72. Patel SS, Neylan JH, Bavaro K, Chai PR, Goralnick E, Erickson TB. Chemical, biological, radiological, nuclear, and explosives (CBRNEs) preparedness for sporting event mass gatherings: A systematic review of the literature. Am J Disaster Med. 2022 Winter;17(1):57–74.
73. Verhelst HM, Stannat AW, Mecacci G. Machine learning against terrorism: how big data collection and analysis influences the privacy-security dilemma. Sci Eng Ethics. 2020;26(6):2975–84.

It's Contagious

10

Ramesses A. Akamefula, Matthew J. Kraeutler, and Mary K. Mulcahey

10.1 Introduction

Physicians often prioritize event coverage, managing overuse injuries and treating acute issues such as fractures, sprains, and strains when tending to athletes; however, infectious diseases rank among athletes' most common health concerns [1]. Participation in organized sports can facilitate the spread of communicable diseases through shared activities, equipment, living and training spaces, and travel to high-risk areas. Recent pandemics and infectious disease outbreaks at multisport events have increased awareness of this risk among most athletes and clinicians. Risks such as sharing bottles, exercise equipment, and close contact during competition increase the susceptibility of athletes to infections compared to the general population [2]. Contact sports further elevate the risk by increasing exposure to open wounds, body fluids, and potential infections [3]. All team physicians should be prepared to offer appropriate advice and respond to infectious exposures to athletes, aiming to protect them from communicable diseases and minimize the likelihood of outbreaks. Some of the most common infections that sports medicine physicians must be prepared to face are the following:

- *Staphylococcus aureus* (including methicillin-resistant and methicillin-sensitive *Staphylococcus aureus*)
- Herpes simplex virus (HSV)
- Fungal infections (e.g., tinea corporis, tinea pedis, and tinea cruris)
- Coronavirus and other community-acquired respiratory viruses
- Mononucleosis
- Conjunctivitis (pink eye)

10.2 Case Anecdote/Challenge

Imagine a scenario where a renowned soccer team is preparing for the World Cup finals. The team has trained rigorously for months, and their star player, Alex, is in peak physical condition. Just days before the finals, Alex starts feeling unwell with a high fever and severe fatigue. The team's medical staff quickly diagnoses him with influenza, which he contracted while traveling for a previous match. Despite their best efforts, Alex's performance is significantly hindered, and the team loses a crucial match. This incident

R. A. Akamefula
Tulane University School of Medicine, New Orleans, LA, USA

M. J. Kraeutler
Department of Orthopaedic Surgery & Rehabilitation, Texas Tech University Health Sciences Center, Lubbock, TX, USA

M. K. Mulcahey (✉)
Department of Orthopaedic Surgery and Rehabilitation, Loyola University Medical Center, Maywood, IL, USA

M. R. Hutchinson, S. Rocha Piedade (eds.), *Sports Medicine: It's Environmental*, https://doi.org/10.1007/978-3-031-84493-5_10

not only affects the team's chances of winning but also raises concerns about the health and safety of the other players and staff who were in close contact with him. This scenario underscores the critical importance of infectious disease prevention in sports medicine. Athletes are constantly traveling, often to regions with different health risks, and they interact closely with many people, increasing their vulnerability to infections. The case of Alex highlights how a seemingly common illness like the flu can have significant repercussions, not just for individual athletes, but for entire teams and events. It emphasizes the need for stringent vaccination protocols, regular health monitoring, and prompt medical interventions to ensure that athletes remain healthy and perform at their best, safeguarding both their careers and the competitive integrity of sports.

10.3 Epidemiology and Lessons Learned from Prior Outbreaks

10.3.1 COVID-19 Pandemic

The declaration of COVID-19 as a pandemic by the World Health Organization on March 11, 2020, marked a significant global health crisis, with over 14 million confirmed cases and nearly 600,000 deaths by July 19, 2020. SARS-CoV-2 spreads rapidly through respiratory droplets and surface contact, which led governments to enforce strict stay-at-home orders to mitigate transmission. The sports world swiftly responded with cancellations and postponements of events at all levels, recognizing the risk posed by close contact among athletes, coaches, and fans. Youth sports were similarly affected, with seasons abruptly terminated and activities restricted. Past epidemics, including the 2003 SARS outbreak, offer some context. However, the previous scale of disruption caused by COVID-19 necessitated a concerted effort to establish consistent safety measures for the resumption of sports activities, prioritizing the health and well-being of all involved.

10.3.2 Zika Virus and the 2016 Rio Olympic Games

While historically considered mild, the *Zika virus* gained attention after outbreaks in Micronesia in 2007 and French Polynesia in 2013 [4, 5]. By 2015, it reached Brazil, leading to significant concerns due to its association with microcephaly and Guillain–Barre syndrome [6, 7]. During Brazil's preparation for the 2016 Olympic Games, the World Health Organization (WHO) declared a public health emergency of international concern due to Zika's potential impact. Despite reassurances of low risk during the Games, concerns persisted, prompting proactive measures by organizations such as the New Zealand Olympic Committee (NZOC) to safeguard athletes' health. The evolving nature of Zika risk mandated a systematic approach, including post-Rio Zika screening [8].

Since the emergence of the *Zika virus* threat in December 2015, the NZOC worked closely with High Performance Sport New Zealand and the New Zealand Ministry of Health (MOH) to address the risk at three stages: pre-Games, during the Olympic Games, and post-Games. Despite the anticipated low transmission risk during the Games, there was a strong desire from staff and athletes to establish a post-exposure testing protocol, which prompted a single serological evaluation to be conducted after returning to NZ. After the Rio Olympic Games, Olympic athletes were provided with additional information and opportunities to consult with medical practitioners due to evolving Zika recommendations [9].

The complex clinical landscape surrounding the *Zika virus* in Brazil, coupled with intense international media attention and an evolving understanding of its consequences, led to uncertainty among prospective team members ahead of the Rio Olympic Games. Despite this uncertainty, the NZ team adopted a proactive approach to mitigate risks, educate, and disseminate information, resulting in high satisfaction among team members. While reports claim no Zika cases during the Olympic Games, the asymptomatic nature of many infections makes this difficult to confirm

[9]. Despite this, Zika remains a significant public health issue, necessitating ongoing mosquito avoidance measures and education for traveling athletes to mitigate risks effectively, particularly regarding pregnancy outcomes.

10.3.3 PyeongChang Winter Olympic Games

During the 2018 PyeongChang Winter Olympics, the Wonju Severance Christian Hospital (WSCH) operated as the emergency department at the Pyeongchang Mountain Polyclinic, catering to staff and athletes from various nations. The Pyeongchang polyclinic's emergency department (ED) was staffed with employees primarily from WSCH and Severance Hospital in Seoul. To ensure continuous coverage, work shifts were organized, with two trauma surgeons alternating every two days, three emergency physicians rotating every three days, and three orthopedic surgeons assigned to each 24-hour shift. Seven nurses followed a similar three-shift system, while four emergency medical technicians worked 24-hour shifts on a four-day rotation [10].

The respiratory tract was the primary system affected in patients visiting the ED, followed by the gastrointestinal tract. Cold weather and sudden environmental changes could have exacerbated these conditions, leading to a higher prevalence of cases. This pattern of high respiratory infection rates aligns with findings from other elite sporting events and is commonly associated with winter rather than summer events [11, 12]. Patient numbers typically rose toward the conclusion of the Olympic Games, likely due to increased competition schedules, a trend that mirrored observations from the 2002 Salt Lake City Winter Olympic Games [13].

The polyclinic encountered specific viral infections, namely, influenza and norovirus. While influenza rapid test kits were accessible daily in the ED, Tamiflu and PeramiFlu injections were readily prepared. Norovirus was confirmed in 168 staff members staying at accommodations near Pyeongchang during the Olympics. To prevent norovirus from spreading to the Olympic Village, infection control measures were implemented, which included immediate response by infection control personnel and distribution of treatment guidelines. Suspected patients were isolated at their accommodations, and "norovirus busters" (staff assigned to patients possibly affected with norovirus) visited athletes to assess conditions, collect specimens, and provide medication. Despite these efforts, a Swiss athlete participating in skiing became the first confirmed norovirus case in the athlete's village. However, proactive measures and strict isolation protocols limited further transmission, with only four confirmed cases ultimately recorded at the Olympic Village. Ultimately, sports medicine physicians should continue to educate themselves on managing unexpected issues like norovirus infections to aid in shaping medical services for the upcoming Olympic Winter Games and other international athletic events.

10.3.4 Preparedness and Education

Orthopedic surgeons must remain adaptable and responsive amid the evolving landscape of pandemics, especially regarding inpatient surgical selection and patient management. Adjusting surgical priorities is critical to meet the demands of the broader healthcare system. Surgical selection priorities may vary depending on the local community's needs and the phase of the infectious outbreak in a particular area. Maintaining personnel safety and mitigating virus transmission are paramount for orthopedic programs to continue essential functions. Emerging technologies like inpatient telemedicine and online file-sharing applications can facilitate continued operations while safeguarding medical staff and patients from infectious agent transmission. Ultimately, surgeons should adhere to the Centers for Disease Control and Prevention guidelines, utilize personal protective equipment (PPE) when appropriate, establish teams that practice inpatient physical/social distancing, employ online file sharing for clinical communication, understand the department's surgical selection policy, and prioritize routines that promote physician well-being [14].

The COVID-19 pandemic specifically prompted significant changes in postgraduate medical and surgical training, particularly affecting orthopedic sports medicine fellows. Recommendations to delay elective procedures led to reduced case volumes, impacting fellows' surgical exposure, which is critical for independent practice preparation [15]. In 2018, Perrone et al. [16] performed a cross-sectional study investigating the impact of COVID-19 on sports medicine fellowship classes in 2020. This group surveyed orthopedic sports medicine fellows in April 2020 revealing widespread implementation of volume-limiting policies, with many fellows experiencing substantial decreases in weekly case volumes [16]. To compensate for the lack of operative experience and case volume, some sports medicine fellowships adopted web-based educational initiatives, including virtual didactics and remote journal clubs [16]. Although concerns initially arose regarding the impact on fellows' training, program directors (PDs) expressed confidence in fellows' future patient care abilities. Moreover, PDs reported adapting to new educational methods and recognized fellows' improved proficiency in virtual medicine [17].

Alongside modifications that may be seen within the operating room, the COVID-19 pandemic also significantly impacted sports medicine fellowship training in terms of missed opportunities outside the operating room. Interacting with athletes of various skill levels and developing relationships with their support network are crucial aspects of fellowship training [18]. However, the widespread cancellation of sporting events in 2020 deprived fellows of these invaluable experiences. A survey of sports medicine fellows revealed that nearly half considered this loss of sports coverage opportunities to be the most significant setback to their education [19]. While fellows previously had ample exposure to sports coverage, this drastically decreased during the pandemic. With athletic activities back to normal now, sports fellows can once again benefit from the comprehensive training and exposure essential for their professional development.

10.3.5 Innovations in Infection Control

The COVID-19 pandemic accelerated the adoption of telemedicine for clinical care, including orthopedics. Orthopedic surgeons may opt to explore telemedicine options, especially during a pandemic or when an athlete is isolated due to an infection. While telemedicine may not fully replicate in-person exams, systematic virtual examinations can aid in triaging and managing common musculoskeletal conditions, especially when performing outpatient injury evaluations [20]. Before the COVID-19 pandemic, researchers investigated telemedicine's effectiveness in orthopedic consultations, and the results showed promising outcomes. For instance, in 2016, Buvik et al. [21] performed a study that compared video consultations with standard visits for various orthopedic conditions, finding it safe and satisfactory for select patients. Although telemedicine has successfully treated orthopedic trauma and postoperative rehabilitation, its application in sports medicine diagnoses remains relatively unexplored [20, 22–24].

10.4 Epidemiology, Management, and Treatment of Common Contagions in Sport

10.4.1 *Staphylococcus aureus* (Methicillin-Resistant *Staphylococcus aureus* (MRSA) and Methicillin-Sensitive *Staphylococcus aureus* (MSSA))

Historically, MRSA was primarily associated with healthcare settings, but the emergence of community-associated MRSA (CA-MRSA) has led to an increase in infections among otherwise healthy individuals, including athletes. The prevalence of CA-MRSA infections among athletes, especially in contact sports such as football and wrestling, is noteworthy. A 2016 study demonstrated that CA-MRSA infections affect approximately 0.6% of high school football players and 0.9% of high school wrestlers each season [25]. This risk escalates significantly among collegiate wrestlers, with a staggering 76% found to be carriers of CA-MRSA, a proportion notably higher than in other sports [26]. Ultimately, direct contact with colonized or infected athletes appears to be the primary origin of these infections [27]. Of the many skin infections that CA-MRSA may cause (e.g., impetigo, erysipelas, cellulitis, folliculitis, furuncles, carbuncles, and abscesses), impetigo causing blisters, open sores, and/or honey-colored crusts stands out as the most prevalent (Fig. 10.1) [28].

The transmission dynamics of CA-MRSA are multifaceted, with direct contact between colonized or infected athletes being the primary mode of spread. This underscores the importance of stringent hygiene practices and infection control measures within sports environments. Although it is common to use antistaphylococcal cleaning supplies and equipment, the efficacy of this practice remains unclear [29]. Nonetheless, evidence suggests that implementing a written policy detailing protocols for equipment sanitation, identification, and prompt treatment could potentially mitigate the occurrence of CA-MRSA skin infections [30].

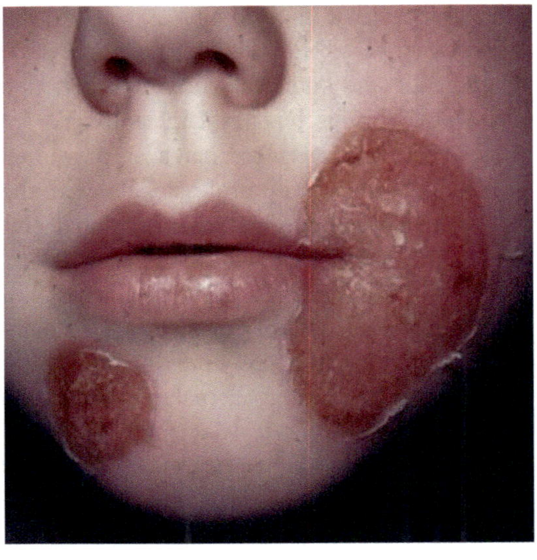

Fig. 10.1 Impetigo infection of the face caused by *Staphylococcus aureus* [31]. (Reprinted with permission from Elsevier)

Decolonization strategies aimed at eradicating MRSA carriage in individuals with recurrent infections or during outbreaks present a controversial yet potentially beneficial intervention. The preferred decolonization approach involves chlorhexidine washes combined with mupirocin ointment application two times per day for seven days [32, 33]. However, the efficacy of this practice in athletes has not been studied extensively, highlighting the need for further research [34]. The emergence of CA-MRSA has introduced complexities into the empiric treatment of bacterial skin infections in recent decades [35]. Moreover, discerning CA-MRSA from MSSA is challenging as there are no distinct signs or symptoms for differentiation [36].

A comprehensive approach involving topical and systemic treatments is crucial in treating MRSA skin infections among athletes. For nonpurulent MRSA infections, either trimethoprim-sulfamethoxazole (160 mg/800 mg) orally twice daily for 7–14 days or clindamycin 300–450 mg orally given three times per day for 7–14 days, in combination with topical mupirocin ointment applied directly to the lesion can be recommended as initial treatment [37, 38]. Culturing infections to determine antibiotic sensitivity is also essential for optimizing treatment outcomes [38]. Additionally, formal irrigation and debridement may be indicated and performed at the bedside if an abscess is present or the infection is purulent, with the addition of antibiotic therapy considered for more severe infections that may not be solely handled with irrigation and debridement [36]. Serious infections characterized by widespread skin involvement or systemic symptoms necessitate treatment with intravenous antibiotics, typically administered within a hospital setting [39].

10.4.2 Herpes Simplex Virus (HSV)

HSV types 1 and 2 are common infectious agents within contact sports, especially in wrestling. In

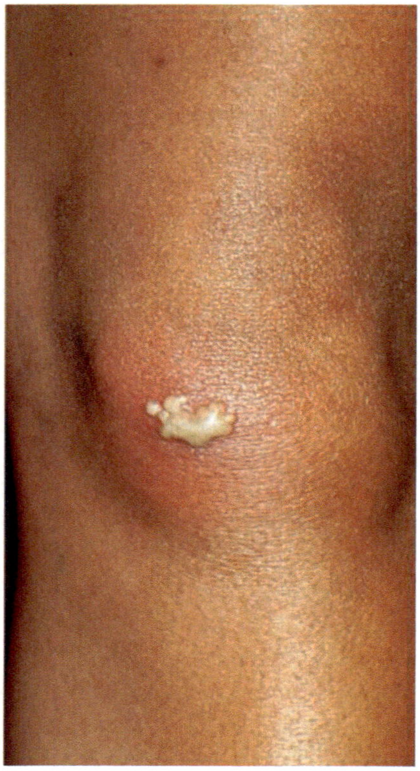

Fig. 10.2 Herpes gladiatorum infection of the knee [42]. (Reprinted with permission from Elsevier)

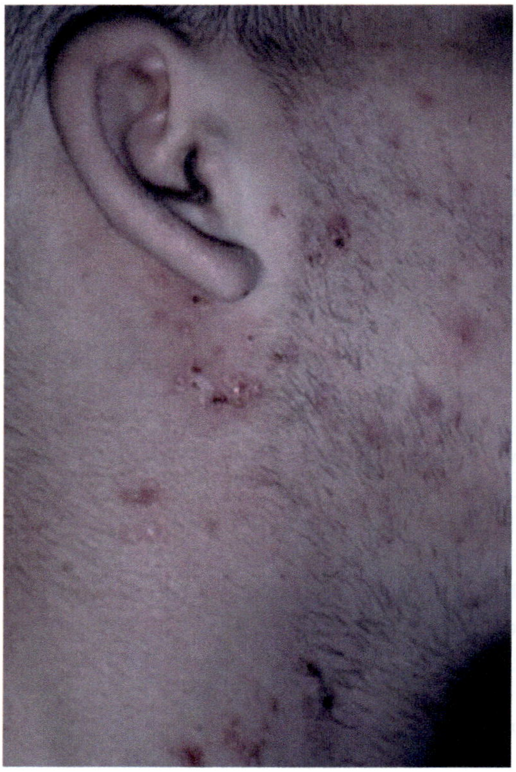

Fig. 10.3 Herpes simplex virus infection of the face [31]. (Reprinted with permission from Elsevier)

1964, Selling and Kubrick originally termed the infectious agent with respect to wrestling as herpes gladiatorum (HG) [40]. HG became further well known after reported outbreaks such as in 1989 after a wrestling camp in Minnesota reported that 60 out of 175 wrestlers contracted the virus, causing the Public Health Department to close the camp [40]. Between 20% and 40% of college wrestlers become infected with HG annually [41]. Clinically, HSV skin infections may be described as diffuse patches of small vesicles ranging from 1 to 2 mm in diameter with a slightly reddened base (Fig. 10.2).

In 2018, Peterson et al. [28] performed a clinical review looking at infectious diseases within contact sports and reported that 94% to 97% of HSV infections among wrestlers are via HSV type 1. The infection is almost entirely transmitted through skin-to-skin contact, with previous studies showing that training mats are not a significant contributor to the spread of HSV type 1 [43]. Beyond this, it is predicted that between 29% and 30% of all high school wrestlers are either colonized or infected with HSV. However, only around 3% of high school wrestlers become symptomatic and know they carry the infection [44]. Due to this discrepancy, large outbreaks are prone to occur. Furthermore, diagnosis may often be missed by healthcare professionals, with the correct diagnosis being made at the initial presentation in less than 10% of cases [45].

Typically, 90–93% of HSV symptoms will occur within eight days of exposure [43]. Primary HSV outbreaks may initially present with symptoms including sore throat, malaise, fever, and anterior/posterior cervical adenopathy [46]. After, 1- to 2-mm diameter patches of vesicles with an erythematous base may coalesce, with multiple dermatomes and extremities being affected (Fig. 10.3). Recurrent HSV infection usually involves fewer vesicles with a shorter duration of physical symptoms and in the same peripheral nerve or dermatomal distribution as the initial infection.

Due to possible confusion with other skin infections among wrestlers, microbiological testing for HSV types 1 and 2 is strongly recommended when clinical suspicion is high [47]. Viral culture and HSV polymerase chain reaction (PCR) are reasonable forms of diagnostic testing due to their excellent sensitivity and exceptional specificity [48], However, despite its higher cost than viral culture, PCR typically provides more rapid results [28, 49].

Athletes should be treated with oral valacyclovir 500 mg or 1 g once per day when clinical suspicion is high [50]. Those infected are then to be withheld from contact with all equipment and other athletes and personnel until the infection has cleared [51]. Symptoms from HSV infection may last up to 10–14 days, even with treatment [39]. However, medication may be able to shorten symptom duration to 2–5 days, allowing an earlier return to sport (RTS) [50].

10.4.3 Fungal Infections

Also known as "ringworm," tinea corporis affects many athletes participating in contact sports, especially wrestlers. Up to 60% are infected each season [46]. Tinea corporis almost exclusively spreads by direct skin-to-skin contact, with open sore lesions increasing the risk of transmission between athletes [52]. Symptoms most often appear around 3–5 days after exposure and typically initiate as a small, red-colored spot that expands into a circular annular lesion (Figs. 10.4 and 10.5). In many cases, the peripheral border of the lesion is described as flaky and red, while the center is more transparent in color.

Diagnosis of tinea corporis is typically made based on physical examination findings. However, potassium hydroxide (KOH)-prepped skin scrapings are highly sensitive and specific for tinea corporis and may provide enhanced screening and diagnosis [55].

Treatment prior to confirmation KOH testing may result in false-negative results. In this case, it may be advised to perform a fungal culture to confirm a negative KOH test result. It should be noted, however, that fungal culture may take up to 21 days before results are available, so empiric treatment of skin or hair using antifungal agents is recommended.

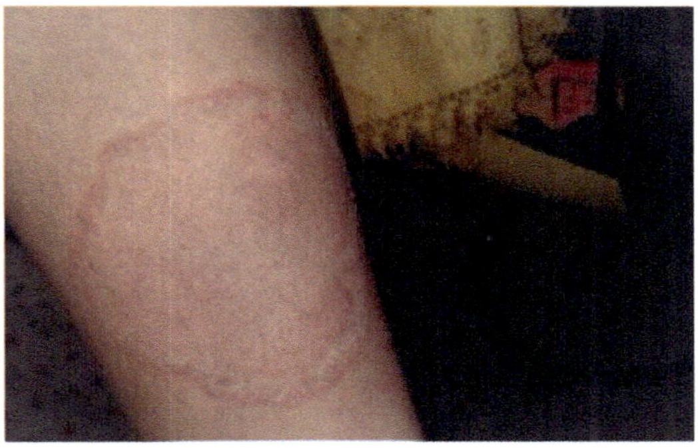

Fig. 10.4 Tinea corporis infection of the lower extremity [53]. (Reprinted with permission from Elsevier)

Treatment of tinea corporis is based on a delicate balance between infection control, risk of transmission, and the interests and goals of the athletes. Due to the extensive period it can take for tinea corporis to heal fully (21 days of treatment), guidelines have been proposed that allow wrestlers and athletes to participate in contact sports during treatment. More specifically, the National Collegiate Athletics Association (NCAA) has allowed participation among athletes currently infected with tinea corporis as long as the athlete has been treated with oral or topical treatment for 72+ hours on the skin and 14 days on the scalp.

To keep fungal infections under control, athletes participating in contact sports should focus on adequate hygiene and routine skin checks. When recurrent tinea corporis occurs due to a non-visible lesion, oral antifungal medications have been shown to decrease possible recurrent infections [56].

10.4.4 Community-Acquired Respiratory Viruses

Athletes are highly susceptible to acquiring acute respiratory infections (ARIs) during winter sports events, often coinciding with peak viral activity

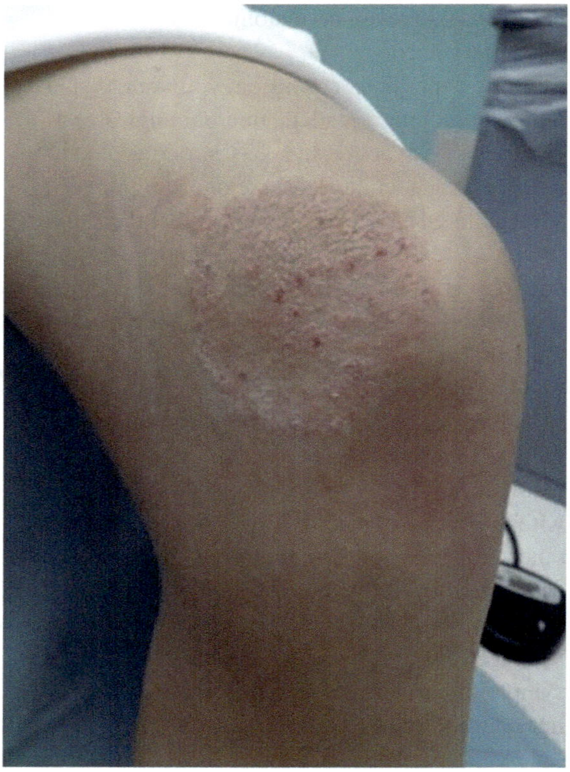

Fig. 10.5 Tinea corporis infection of the knee of a wrestling team cheerleader [54]. (Reprinted with permission from Elsevier)

within the community [57]. Various behavioral factors exhibited by athletes during these events, including utilizing public transportation, congregating in crowded spaces, sharing group accommodations, and engaging in close social interactions, collectively contribute to their increased vulnerability to ARIs [58].

Diagnosing respiratory diseases in athletes requires a comprehensive understanding of their symptoms and potential triggers. Symptoms such as sore throat, cough, nasal congestion, and fever are common indicators of ARIs [59]. Clinicians must recognize these signs, especially during major sporting events where athletes are at a higher risk of exposure to respiratory viruses.

Furthermore, diagnostic testing plays a crucial role in confirming respiratory infections. Nasal swab specimens and polymerase chain reaction (PCR) assays commonly detect viral pathogens, including influenza, rhinovirus, respiratory syncytial virus, and coronaviruses. Timely and accurate diagnosis enables prompt intervention and prevents the spread of infections among athletes and staff.

Effective treatment of respiratory diseases in athletes involves a combination of pharmacological and non-pharmacological interventions. Antiviral medications may be prescribed to mitigate the severity and duration of viral infections, while symptomatic relief medications help alleviate discomfort and improve respiratory function. However, it is essential to note that rest and hydration are fundamental components of treatment, allowing the body to recover and combat the infection effectively. Athletes may need to adjust their training schedules and workload to accommodate the healing process and prevent exacerbation of symptoms.

Preventing respiratory diseases in athletes requires a proactive approach addressing individual and environmental factors. Implementing strict hygiene measures, such as frequent handwashing, respiratory etiquette, and avoiding close contact with sick individuals, is paramount in reducing the risk of infection transmission. Moreover, vaccination against influenza and other respiratory viruses can protect athletes, minimizing the likelihood of contracting viral illnesses during competitions and training sessions. Adequate nutrition, sufficient sleep, and stress management support the immune system and enhance resistance to respiratory infections.

10.4.5 Infectious Mononucleosis (IM)

No evidence indicates that IM is more or less prevalent among student-athletes than nonstudent-athletes. However, IM appears to be more common in populations living in close quarters, such as college students in dorms and active military personnel, and it ranks as the top cause of lost time for new Army recruits [60, 61]. Management of IM in athletes involves supportive care, with caution regarding splenic rupture risks. Return-to-play decisions are complex and individualized, considering age, severity of illness, sport type, season timing, and psychological readiness. There is a need for shared decision-making (SDM) between healthcare providers, athletes, and their support network. While evidence suggests refraining from high-exertion activities for at least two weeks and avoiding contact sports for 3–4 weeks, decisions should be tailored to each athlete's clinical status and preferences [62].

10.4.6 Conjunctivitis (Pink Eye)

Viral and bacterial infections of the conjunctiva are common among both athletes and nonathletes. Viral and bacterial conjunctivitis can be contracted directly from infected individuals or can result from the abnormal proliferation of native conjunctival flora [63]. Common routes of transmission include contaminated fingers and contaminated fomites [64, 65]. Although most of these infections are benign and self-limiting, they can be easily transmitted among contact sport athletes. For example, in 2003, Martin et al. [66] performed a study looking at demographics of a specific outbreak of conjunctivitis on a college campus. This group discovered that during this outbreak of *Streptococcus pneumoniae* conjunctivitis, athletes were found to be at a higher risk of infection compared to nonathletes. However, methods for controlling such outbreaks remain controversial and insufficiently studied. Currently, the NCAA does not mandate treatment or proof of resolution before athletes return to practice or competition when infected with conjunctivitis.

Regarding treatment, most cases of bacterial conjunctivitis resolve on their own and do not require treatment if they are uncomplicated. However, conjunctivitis caused by gonorrhea or chlamydia, as well as conjunctivitis in contact lens wearers, should be treated with antibiotics [67]. Moreover, topical ophthalmic antibiotics may shorten the duration of bacterial conjunctivitis and may help reduce transmission [68]. Viral conjunctivitis is typically managed with supportive care. Allergic conjunctivitis symptoms can be relieved with avoiding the offending agent, using saline solution or artificial tears to physically dilute and remove the allergens, antihistamines, or mast cell inhibitors [69, 70].

10.5 Prevention and Risk Reduction

While infectious agents such as COVID-19 may manifest as mild symptoms in most athletes, implementing prevention strategies is imperative for several reasons. Foremost among these is the need to curb the transmission of infectious agents, thereby mitigating the risk of spreading it to other athletes, coaches, and spectators, particularly in older individuals and those with compromised immune systems. Additionally, preventing contagious agents is crucial for athletes in maintaining uninterrupted training regimens and safeguarding against potential impacts on their respiratory health and aerobic capacity, both in the short and long term. Sports medicine providers can aid athletes and teams in preventing the spread of infectious agents by advocating the following preventive measures (summary of measures provided below in Table 10.1).

Table 10.1 Prevention summary chart

Prevention measures	Summary
Hand hygiene	Frequent handwashing with soap and water or using hand sanitizer with at least 60% alcohol is recommended. Cleaning and disinfecting frequently touched objects and surfaces are essential, especially considering the prolonged survival of viruses such as COVID-19 on surfaces
Social distancing	Social distancing involves maintaining physical distance from others, though specific recommendations vary. Strategies such as staggered or side-by-side running can mitigate transmission risks during sports activities. Enhanced ventilation and maximizing open spaces are advised for indoor sports to reduce infection risks, especially during known outbreaks
Travel considerations	During pandemics, minimize travel. Countries may implement various restrictions, such as halting flights, entry limitations, and mandatory quarantine upon returning. Entry screenings, including temperature checks and symptom assessments, may be conducted. Athletes should adhere to these measures and avoid unnecessary travel to minimize the risk of spreading infectious agents
Face coverings/ masks	Use facial masks in suspected community-acquired respiratory infections, but their efficacy for healthy athletes is uncertain. Misuse of masks can strain supply and demand. Specific concerns include respiratory difficulties, potential microbial growth, and impediments to ventilation during exercise. Masks are recommended for coaches, trainers, and allied health staff during practices and meetings but may hinder athletic performance during exercise
Training modifications	Intense training may temporarily suppress the immune system, increasing susceptibility to infection. Detraining effects due to inactivity include reduced cardiovascular fitness and muscle weakness. Gradual reconditioning programs are recommended to counteract detraining effects, focusing on diverse home-based activities initially. Individualized rehabilitation protocols should be integrated into athletic programs to facilitate a safe return to sport
Immunizations/ vaccinations	Athletes should ensure routine vaccinations are up to date and consider additional vaccines based on individual circumstances, including travel and immunocompromised status. Vaccine timing should allow for temporary effects to dissipate before sports activity begins, ensuring an effective immune response. Special considerations may include alternate injection sites and managing systemic reactions to prevent disruption to performance

10.5.1 Hand Hygiene

Common recommendations involve frequent handwashing with soap and water for at least 20 s or utilizing hand sanitizer containing at least 60% alcohol when soap and water are inaccessible. Given some viruses' ability to persist on surfaces for extended periods, it is essential to routinely clean and disinfect frequently touched objects and surfaces [71].

10.5.2 Social Distancing

The Centers for Disease Control and Prevention (CDC) describes social distancing as putting physical distance between yourself and others when possible, assuming that viral particles dissipate or settle before reaching others. Because transmission depends on a multitude of factors, there is currently no single distance recommendation [72]. This practice was strongly advocated for during the COVID-19 pandemic and promoted by professional athletes as well. Blocken et al. [73] conducted a study that suggested that the risk of exposure varies depending on the arrangement of runners, with inline running posing the highest risk. As athletes resume training, strategies such as staggered or side-by-side running may mitigate transmission risks. While evidence on SARS-CoV-2 aerosolization is still emerging, precautions for indoor sports should be heightened during times of known infectious outbreak due to increased infection risks, possibly including enhanced ventilation and maximizing open spaces [74].

10.5.3 Travel Considerations

During the COVID-19 pandemic, numerous countries implemented travel limitations and restrictions to mitigate transmission rates of infectious agents. These measures varied from halting flights and limiting entry for travelers from affected regions to mandating self-isolation quarantines upon returning from specific destinations. Additionally, countries conducted entry screenings, such as temperature checks and symptom assessments for COVID-19 [75]. During a pandemic, all travel for athletes should be minimized.

10.5.4 Face Coverings and Masks

High suspicion of community-acquired respiratory agents prompts immediate use of facial masks to reduce spread and transmission [76]. However, during the COVID-19 pandemic, athletes who showed no symptoms of illness were not recommended to wear masks to prevent COVID-19 infection in community settings or during travel, as it did not notably decrease the risk of contracting the virus. During times of pandemics, misuse of masks can strain the supply and demand, potentially leaving healthcare workers and those infected without adequate personal protective equipment (PPE). Moreover, wearing masks during training may be impractical for athletes [75]. Studies further suggest the varying effectiveness of different types of masks, with cloth masks potentially offering less protection than surgical masks. Specific to athletes, an increased flow rate during exercise may further diminish the effectiveness of surgical masks [74, 77]. The efficacy of masks for healthy athletes remains uncertain, with concerns regarding respiratory difficulties and potential microbial growth due to sweat. Mask usage for healthy athletes may also further impede an athlete's ability to ventilate during exercise due to decreased tidal volume and increased carbon dioxide retention [74, 78]. However, athletes, coaches, trainers, and allied health staff should still be encouraged to wear masks during practices and meetings. In addition, athletes should wear masks during one-on-one interactions when not exercising [79].

10.5.5 Training Modifications and Return to Sport (RTS)

It has been suggested that prolonged and intense training may temporarily suppress the immune system, lasting from hours to days. However, the validity of this "open window" theory, which suggests increased susceptibility to infection following vigorous exercise, remains uncertain [80]. Should an athlete be sidelined secondary to infectious causes, it should be noted that detraining effects due to inactivity, including muscle weakness, reduced cardiovascular fitness, and metabolic changes, are well documented. This period of inactivity leads to decreased muscle size, altered fiber composition, and diminished cardiac and respiratory function [81–84]. To counteract these effects of deconditioning, gradual reconditioning programs are recommended by organizations such as the National Strength and Conditioning Association and Collegiate Strength and Conditioning Coaches Association. These guidelines aimed to prevent exertional illnesses such as heat exhaustion during the transition back to full activity. Initially, athletes are encouraged to engage in diverse home-based activities to mitigate detraining effects. As restrictions ease, individualized and group rehabilitation protocols can be integrated into athletic programs to facilitate sport-specific training [85]. Team physicians play a vital role in developing and implementing safety protocols and return-to-play protocols, as well as educating teams on their importance. By addressing concerns and disseminating evidence-based information, physicians can guide athletes through a safe return to play. However, it should be emphasized that RTS protocols ultimately depend on the individual athlete, infectious agent, and rules implemented by the local health guidelines.

10.5.6 Immunizations and Vaccinations

When considering the immunization needs of athletes, it is essential to take a comprehensive approach. Immunization considerations involve ensuring that the athlete's routine vaccinations are up to date as per national recommendations and assessing any unique circumstances they may face. Athletes who are immunocompromised or traveling internationally may require additional vaccinations. Vaccine recommendations can vary not just between countries but also within regions. Factors such as injection site, potential localized reactions, and timing in relation to sports participation must be carefully considered. Managing systemic reactions or side effects of vaccination, such as fever and pain, is also crucial to prevent disruption to an athlete's performance. Clinicians may opt for alternate injection sites, particularly for athletes in throwing sports, to minimize interference with training and competition. Athletes should ultimately be immunized before a competitive event or season starts. This timing allows for the acquisition and administration of necessary vaccines well in advance, ensuring any temporary effects of the vaccine dissipate before sports activity begins and allowing sufficient time for an effective immune response to develop [86, 87].

The 2016 Olympic Summer Games in Brazil heightened global attention not only due to the event's prominence but also due to infection risks, particularly highlighted by the *Zika virus* outbreak. This situation led to an increased emphasis on preventive measures, including vaccinations, with the World Health Organization providing specific health advice and vaccine information for travelers, a step up from the general health measures advised during the 2012 London Olympics. In 2017, Tafuri et al. [88] performed a study that underscored the heightened risk of infectious disease transmission among athletes due to factors like crowded living conditions and international travel. Despite limited research on athlete-specific vaccination strategies, this group highlighted the importance of immunizations tailored to the unique needs and risks of athletes, advocating for rigorous, evidence-based approaches to their vaccination protocols.

Athletes should receive all age-appropriate vaccinations recommended for the general population, with additional immunizations tailored for international travel, such as hepatitis A, typhoid fever, yellow fever, and meningococcal disease, depending on the region [88]. Annual influenza vaccination is advised due to the high risk of transmission in crowded environments and the critical need to maintain health for optimal performance. Routine immunizations, including seasonal flu vaccines and decennial booster doses for tetanus and diphtheria, are essential. Specific and robust recommendations have been made for tetanus vaccinations, especially for athletes who may come into contact with soil contaminated with *C. tetani* and are at risk of injuries, such as in horse racing and cross-country. Additionally, hepatitis B vaccinations are strongly advised for sports where there is a high risk of exposure to blood and bodily fluids, including football, rugby, boxing, wrestling, soccer, and ice hockey [87, 89]. Moreover, athletes are encouraged to serve as role models to promote vaccination acceptance within their communities, with immunization extended to all staff members, including coaches, referees, and trainers, to enhance overall coverage. Close collaboration between infectious disease specialists, sports medical staff, and sport federations is crucial to implement effective immunization strategies and address the specific needs of athletes.

Extreme sports events held in remote or challenging environments pose additional health risks due to poor sanitation and limited access to healthcare. Athletes competing in these events may face increased susceptibility to infections from environmental conditions and vector exposure [1].

10.6 Conclusion: Achieving Success Through Health

To effectively maintain athletes' health amid infectious challenges, sports medicine providers must adopt comprehensive strategies encompassing risk reduction practices, training modifications, and preparedness measures. Hand hygiene emerges as a cornerstone, advocating frequent washing with soap and water or alcohol-based sanitizers and regular cleaning of shared surfaces. Social distancing remains pivotal, and travel considerations demand careful navigation that urges minimal travel during pandemics while adhering to screening measures upon entry. Despite debates over the efficacy of masks for healthy athletes during exercise, their use by coaches and staff is encouraged, especially indoors, balancing potential benefits against practical limitations. Training modifications, immunizations, and gradual reconditioning programs are essential for athletes sidelined by infections, minimizing detraining effects and ensuring a safe return to play under physicians' guidance. The lessons learned from past infectious threats, such as the *Zika virus* outbreak during the 2016 Rio Olympic Games, underscore the importance of proactive measures and education dissemination. As we progress, innovations such as telemedicine offer promising remote care and diagnosis avenues, particularly during pandemics or when athletes are isolated.

Overall, a multifaceted approach combining preventive measures, education, and innovative solutions is essential to safeguard athletes' health and ensure the uninterrupted pursuit of their athletic endeavors in the face of infectious challenges. Sports medicine providers, coaches, administrators, and athletes must unite collectively to implement and uphold rigorous preventive measures. Fostering a culture of transparency and open communication is paramount, enabling swift response to emerging threats and the dissemination of evidence-based guidelines. By prioritizing the health and safety of athletes above all else, we can mitigate the impact of infectious challenges and uphold the integrity of sportsmanship and competition.

10.7 Take-Home Pearls

- *Awareness of Specific Organisms:* Be vigilant for infections caused by *Staphylococcus aureus* (including MRSA and MSSA), HSV, and fungal infections. These organisms commonly affect athletes, particularly those engaged in contact sports such as football, wrestling, and others.
- *Prevention Strategies:* Implement stringent hygiene practices and infection control measures within sports environments to mitigate the spread of communicable diseases. Written policies detailing protocols for equipment sanitation, identification, and prompt treatment of skin infections can help reduce infection rates.
- *Risk Reduction Strategies:* Advocate for preventive measures such as hand hygiene, social distancing, immunizations, travel considerations, and face coverings to minimize the risk of infectious diseases among athletes and sports personnel.
- *Training Modifications and Return to Sport:* Develop gradual reconditioning programs for athletes recovering from infectious illnesses to prevent exertional illnesses and facilitate safe return to sport. Individualized protocols should consider the athlete's clinical status, sport type, and local health guidelines.

References

1. Harris MD. Infectious disease in athletes. Curr Sports Med Rep. 2011;10(2):84–9.
2. Nattiv A, Puffer JC, Green GA. Lifestyles and health risks of collegiate athletes: a multi-center study. Clin J Sport Med. 1997;7(4):262–72.
3. Luke A, d'Hemecourt P. Prevention of infectious diseases in athletes. Clin Sports Med. 2007;26(3):321–44.
4. Cao-Lormeau V-M, et al. Zika virus, French Polynesia, South pacific, 2013. Emerg Infect Dis. 2014;20(6):1085.
5. Duffy MR, et al. Zika virus outbreak on Yap Island, Federated States of Micronesia. N Engl J Med. 2009;360(24):2536–43.

6. Broutet N, et al. Zika virus as a cause of neurologic disorders. N Engl J Med. 2016;374(16):1506–9.
7. Campos GS, Bandeira AC, Sardi SI. Zika virus outbreak, bahia, Brazil. Emerg Infect Dis. 2015;21(10):1885.
8. Burattini M, et al. Potential exposure to Zika virus for foreign tourists during the 2016 Carnival and Olympic games in Rio de Janeiro, Brazil. Epidemiol Infect. 2016;144(9):1904–6.
9. Hamilton B, et al. Zika virus and the Rio Olympic games. Clin J Sport Med. 2019;29(6):523–6.
10. Kim K, et al. Experiences of the emergency department at the pyeongchang polyclinic during the 2018 PyeongChang Winter Olympic Games. Yonsei Med J. 2019;60(5):474–80.
11. Ruedl G, et al. Sport injuries and illnesses during the first Winter Youth Olympic Games 2012 in Innsbruck, Austria. Br J Sports Med. 2012;46(15):1030–7.
12. Soligard T, et al. Sports injuries and illnesses in the Sochi 2014 Olympic Winter Games. Br J Sports Med. 2015;49(7):441–7.
13. Reeser JC, Willick S, Elstad M. Medical services provided at the Olympic Village polyclinic during the 2002 Salt Lake City winter games. Wmj. 2003;102(4):20–5.
14. Massey PA, et al. Orthopaedic surgical selection and inpatient paradigms during the coronavirus (COVID-19) pandemic. J Am Acad Orthop Surg. 2020;28(11):436–50.
15. Peebles LA, et al. The COVID-19 pandemic increased the use of virtual fellowship interviews, reduced case volume, and limited sideline coverage opportunities during sports medicine surgeon training. Arthrosc Sports Med Rehabil. 2023;5(3):e891–5.
16. Perrone MA, et al. The impact of COVID-19 on the sports medicine fellowship class of 2020. Orthop J Sports Med. 2020;8(7):2325967120939901.
17. Heaps BM, Dugas JR, Limpisvasti O. The impact of COVID-19 on orthopedic surgery fellowship training: a survey of fellowship program directors. HSS J. 2022;18(1):105–9.
18. Peebles LA, et al. Impact of the COVID-19 pandemic on sports medicine patient care. Arthroscopy. 2023;39(2):161–5.
19. Stein CJ, et al. Fellows' perspective of educational changes in medical sports medicine during the COVID-19 pandemic. Cureus. 2022;14(6):e26199.
20. Tanaka MJ, et al. Telemedicine in the era of COVID-19: the virtual Orthopaedic examination. JBJS. 2020;102(12):e57.
21. Buvik A, et al. Quality of care for remote orthopaedic consultations using telemedicine: a randomised controlled trial. BMC Health Serv Res. 2016;16:1–11.
22. Blank E, et al. Early analysis of the United States Army's telemedicine orthopaedic consultation program. J Surg Orthop Adv. 2011;20(1):50.
23. Pastora-Bernal JM, et al. Evidence of benefit of telerehabilitation after orthopedic surgery: a systematic review. J Med Internet Res. 2017;19(4):e142.
24. Prada C, et al. Results of a new telemedicine strategy in traumatology and orthopedics. Telemed e-Health. 2020;26(5):665–70.
25. Braun T, et al. CA-MRSA infection incidence and Care in High School and Intercollegiate Athletics. Med Sci Sports Exerc. 2016;48(8):1530–8.
26. Champion AE, et al. Prevalence and characterization of methicillin-resistant Staphylococcus aureus isolates from healthy university student athletes. Ann Clin Microbiol Antimicrob. 2014;13:33.
27. Kahanov L, et al. Staphylococcus aureus and community-associated methicillin-resistant Staphylococcus aureus (CA-MRSA) in and around therapeutic whirlpools in college athletic training rooms. J Athl Train. 2015;50(4):432–7.
28. Peterson AR, Nash E, Anderson BJ. Infectious disease in contact sports. Sport Health. 2019;11(1):47–58.
29. Fritz SA, et al. Practices and procedures to prevent the transmission of skin and soft tissue infections in high school athletes. J Sch Nurs. 2012;28(5):389–96.
30. Sanders JC. Reducing MRSA infections in college student athletes: implementation of a prevention program. J Community Health Nurs. 2009;26(4):161–72.
31. Carr PC, Cropley TG. Sports dermatology: skin disease in athletes. Clin Sports Med. 2019;38(4):597–618.
32. Watanakunakorn C, et al. The efficacy of mupirocin ointment and chlorhexidine body scrubs in the eradication of nasal carriage of Staphylococcus aureus among patients undergoing long-term hemodialysis. Am J Infect Control. 1992;20(3):138–41.
33. Mody L, et al. Mupirocin-based decolonization of Staphylococcus aureus carriers in residents of 2 long-term care facilities: a randomized, double-blind, placebo-controlled trial. Clin Infect Dis. 2003;37(11):1467–74.
34. Weber K. Community-associated methicillin-resistant Staphylococcus aureus infections in the athlete. Sports Health. 2009;1(5):405–10.
35. Buss BF, et al. Population-based estimates of Methicillin-Resistant Staphylococcus Areus (MRSA) infections among high school athletes—Nebraska, 2006–2008. J Sch Nurs. 2009;25(4):282–91.
36. May TJ, Safranek S. Clinical inquiries. When should you suspect community-acquired MRSA? How should you treat it? J Fam Pract. 2009;58(5):276–8.
37. Cadena J, et al. Dose of trimethoprim-sulfamethoxazole to treat skin and skin structure infections caused by methicillin-resistant Staphylococcus aureus. Antimicrob Agents Chemother. 2011;55(12):5430–2.
38. Stevens DL, et al. Practice guidelines for the diagnosis and management of skin and soft-tissue infections. Clin Infect Dis. 2005;41(10):1373–406.
39. Wilson EK, et al. Cutaneous infections in wrestlers. Sports Health. 2013;5(5):423–37.
40. Belongia EA, et al. An outbreak of herpes gladiatorum at a high-school wrestling camp. N Engl J Med. 1991;325(13):906–10.

41. Agel J, et al. Descriptive epidemiology of collegiate men's wrestling injuries: National Collegiate Athletic Association Injury Surveillance System, 1988–1989 through 2003–2004. J Athl Train. 2007;42(2):303–10.
42. Sagut P, Elston D. Dermatology images: herpes simplex. J Am Acad Dermatol. 2024;91:180.
43. Anderson BJ. The epidemiology and clinical analysis of several outbreaks of herpes gladiatorum. Med Sci Sports Exerc. 2003;35(11):1809–14.
44. Anderson BJ. Prophylactic valacyclovir to prevent outbreaks of primary herpes gladiatorum at a 28-day wrestling camp. Jpn J Infect Dis. 2006;59(1):6–9.
45. Dworkin MS, et al. Endemic spread of herpes simplex virus type 1 among adolescent wrestlers and their coaches. Pediatr Infect Dis J. 1999;18(12):1108–9.
46. Zinder SM, et al. National athletic trainers' association position statement: skin diseases. J Athl Train. 2010;45(4):411–28.
47. Landry GL, Chang CJ. Herpes and tinea in wrestling: managing outbreaks, knowing when to disqualify. Phys Sportsmed. 2004;32(10):34–41.
48. Usatine RP, Tinitigan R. Nongenital herpes simplex virus. Am Fam Physician. 2010;82(9):1075–82.
49. Schmutzhard J, et al. Detection of herpes simplex virus type 1, herpes simplex virus type 2 and varicella-zoster virus in skin lesions. Comparison of real-time PCR, nested PCR and virus isolation. J Clin Virol. 2004;29(2):120–6.
50. Anderson BJ. Valacyclovir to expedite the clearance of recurrent herpes gladiatorum. Clin J Sport Med. 2005;15(5):364–6.
51. Johnson R. Herpes gladiatorum and other skin diseases. Clin Sports Med. 2004;23(3):473–84. x
52. Adams BB. Tinea corporis gladiatorum. J Am Acad Dermatol. 2002;47(2):286–90.
53. Ahmad M, Mobeen A. Efficacy and safety of Majoon Ushba oral and Marham Raal topical in tinea corporis—a randomized open-labeled active-controlled clinical trial. EXPLORE. 2023;19(5):681–8.
54. McClanahan C, Wanat K. Tinea corporis in a wrestling team cheerleader. Int J Women's Dermatol. 2016;2(4):143–4.
55. Kohl TD, Lisney M. Tinea gladiatorum: wrestling's emerging foe. Sports Med. 2000;29(6):439–47.
56. Brickman K, et al. Fluconazole as a prophylactic measure for tinea gladiatorum in high school wrestlers. Clin J Sport Med. 2009;19(5):412–4.
57. Valtonen M, et al. Common cold in team Finland during 2018 winter Olympic games (PyeongChang): epidemiology, diagnosis including molecular point-of-care testing (POCT) and treatment. Br J Sports Med. 2019;53(17):1093–8.
58. Ruuskanen O, et al. Respiratory viral infections in athletes: many unanswered questions. Sports Med. 2022;52(9):2013–21.
59. Puhakka T, et al. The common cold: effects of intranasal fluticasone propionate treatment. J Allergy Clin Immunol. 1998;101(6 Pt 1):726–31.
60. Cohen JI. Vaccine development for Epstein-Barr virus. Human Herpesviruses. 2018:477–93.
61. Williams-Harmon YJ, Jason LA, Katz BZ. Incidence of infectious mononucleosis in universities and US military settings. J Diagn Tech Biomed Anal. 2016;5(1)
62. Putukian M, et al. American Medical Society of Sports Medicine Position Statement: mononucleosis and athletic participation. Clin J Sport Med. 2023;33(4):359–67.
63. Mannis M et al Duanes ophthalmology on CD-ROM. 2006.
64. Høvding G. Acute bacterial conjunctivitis. Acta Ophthalmol. 2008;86(1):5–17.
65. Sattar SA, et al. Spread of acute hemorrhagic conjunctivitis due to enterovirus-70: effect of air temperature and relative humidity on virus survival on fomites. J Med Virol. 1988;25(3):289–96.
66. Martin M, et al. An outbreak of conjunctivitis due to atypical Streptococcus pneumoniae. N Engl J Med. 2003;348(12):1112–21.
67. Azari AA, Barney NP. Conjunctivitis: a systematic review of diagnosis and treatment. JAMA. 2013;310(16):1721–9.
68. Jefferis J, et al. Acute infective conjunctivitis in primary care: who needs antibiotics? An individual patient data meta-analysis. Br J Gen Pract. 2011;61(590):e542–8.
69. Bielory BP, O'Brien TP, Bielory L. Management of seasonal allergic conjunctivitis: guide to therapy. Acta Ophthalmol. 2012;90(5):399–407.
70. Yaylali V, et al. Comparative study of 0.1% olopatadine hydrochloride and 0.5% ketorolac tromethamine in the treatment of seasonal allergic conjunctivitis. Acta Ophthalmol Scand. 2003;81(4):378–82.
71. van Doremalen N, et al. Aerosol and surface stability of SARS-CoV-2 as compared with SARS-CoV-1. N Engl J Med. 2020;382(16):1564–7.
72. Prevention, C.f.D.C.a. About physical distancing and respiratory viruses. [2024 April 16, 2024]; Available from: https://www.cdc.gov/respiratory-viruses/prevention/physical-distancing.html
73. Blocken B, Malizia F, van Druenen T, Marchal T. Towards aerodynamically equivalent COVID19 1.5 m social distancing for walking and running. (n.d.)
74. Mercurio AM, et al. Five myths of COVID-19 for the team physician. HSS J. 2020;16(Suppl 1):173–8.
75. Toresdahl BG, Asif IM. Coronavirus disease 2019 (COVID-19): considerations for the competitive athlete. Sports Health. 2020;12(3):221–4.
76. (NIOSH), T.N.I.f.O.S.a.H. Community respirators and masks 2024 [cited 2024 April16]; Available from: https://www.cdc.gov/niosh/topics/publicppe/community-ppe.html#:~:text=Masks%20also%20filter%20out%20your,from%20germs%20spread%20by%20others.&text=Exposure%20to%20germs%20in%20the,like%20influenza%20and%20COVID%2D19
77. Konda A, et al. Aerosol filtration efficiency of common fabrics used in respiratory cloth masks. ACS Nano. 2020;14(5):6339–47.

78. Lerman Y, et al. External inspiratory resistance of protective respiratory devices: effects on physical performance and respiratory function. Am J Ind Med. 1983;4(6):733–40.
79. Association, N.C.A. Resocialization of Collegiate Sport: Developing Standards for Practice and Competition. 2024 [cited 2024 April 16]; Available from: https://www.ncaa.org/sports/2020/7/14/resocialization-of-collegiate-sport-developing-standards-for-practice-and-competition.aspx
80. Campbell JP, Turner JE. Debunking the myth of exercise-induced immune suppression: redefining the impact of exercise on immunological health across the lifespan. Front Immunol. 2018;9:648.
81. Eirale C, et al. Medical recommendations for home-confined footballers' training during the COVID-19 pandemic: from evidence to practical application. Biol Sport. 2020;37(2):203–7.
82. Mujika I, Padilla S. Detraining: loss of training-induced physiological and performance adaptations. Part I: short term insufficient training stimulus. Sports Med. 2000;30(2):79–87.
83. Phillips SM, McGlory C. CrossTalk proposal: the dominant mechanism causing disuse muscle atrophy is decreased protein synthesis. J Physiol. 2014;592(24):5341–3.
84. Wang Y, Pessin JE. Mechanisms for fiber-type specificity of skeletal muscle atrophy. Curr Opin Clin Nutr Metab Care. 2013;16(3):243–50.
85. Caterisano A, et al. CSCCa and NSCA joint consensus guidelines for transition periods: safe return to training following inactivity. Strength Cond J. 2019;41(3):1–23.
86. Daly P, Gustafson R. Public health recommendations for athletes attending sporting events. Clin J Sport Med. 2011;21(1):67–70.
87. Gärtner BC, Meyer T. Vaccination in elite athletes. Sports Med. 2014;44(10):1361–76.
88. Tafuri S, Sinesi D, Gallone MS. Vaccinations among athletes: evidence and recommendations. Expert Rev Vaccines. 2017;16(9):867–9.
89. Trabacchi V, et al. Immunization practices in athletes. Acta Biomed. 2015;86(2):181–8.

It's Indoors

11

Sujith Kochat and Mark R. Hutchinson

11.1 Introduction

Indoor sporting events offer dynamic, controlled, and reproducible environments for athletes and spectators. Indoor events also introduce a unique set of environmental challenges that can influence performance and injury risk as well as demand careful consideration. These include such issues as lighting, flooring, air quality, proximity of spectators to the playing surface, physical plant and structures, controlled emergency access and egress points, elevated noise exposure, and the advantage of fixed training rooms or emergency facilities. Management is crucial. From logistical hurdles to environmental concerns, the intricacies of organizing and executing these events require a nuanced understanding of both benefits and limitations of indoor venues.

The allure of indoor sports lies in their ability to shield participants from the unpredictability of outdoor conditions. This ensures a more consistent field of play and spectator comfort. Poor lighting, inadequate ventilation, slippery or uncushioned floors, and indoor acoustics can significantly impact performance and place an athlete at increased risk of injury. Crowd management add layers of complexity. Event planning with consideration of injury prevention and execution demand meticulous attention.

S. Kochat (✉)
Advocate Lutheran General Hospital Sports Medicine, Park Ridge, IL, USA
e-mail: sujith.kochat@aah.org

M. R. Hutchinson
UIC Sports Medicine Center, Department of Orthopaedics, University of Illinois at Chicago, Chicago, IL, USA
e-mail: mhutch@uic.edu

Some of these environmental factors can play pivotal role in injury/illness risk but may also be targetable to reduce risk. Indoor air quality (IAQ) significantly influences athletic performance, with suboptimal conditions posing notable challenges to athletes especially those at risk for respiratory illnesses. Poor IAQ, characterized by high levels of pollutants such as volatile organic compounds (VOCs), particulate matter (PM), and carbon dioxide (CO_2), can impair respiratory function and diminish overall physiological efficiency. Exposure to indoor pollutants can lead to respiratory irritation, decreased lung function, and heightened susceptibility to respiratory infections among athletes [1]. These respiratory effects not only compromise aerobic capacity but also impair endurance and recovery rates during intense physical exertion [2]. Furthermore, inadequate ventilation and improper humidity control in indoor environments contribute to discomfort and thermal stress, affecting thermal regulation mechanisms crucial for maintaining optimal body temperature during exertion [3]. The cumulative impact of these IAQ factors underscores the critical need for stringent regulation and monitoring in

M. R. Hutchinson, S. Rocha Piedade (eds.), *Sports Medicine: It's Environmental*,
https://doi.org/10.1007/978-3-031-84493-5_11

indoor sports facilities to mitigate potential performance decrements and safeguard athlete health and well-being. Meticulous planning and innovative solutions help mitigate potential disruptions. They also ensure optimal conditions for competition.

Indoor facilities use more infrastructure to create the controlled environment in which to host these events including lighting, heating, ventilation, and air conditioning. These provide a more reproducible environment for competitors and comfort for spectators but also add to operational expenses. Indoor arenas consume as high as fivefold energy compared to open-air stadiums, due to the principal reason that a constant climate sensitivity must be maintained for both athletes and audiences [4]. An indoor venue requires highly sophisticated lighting systems as well as acoustics management, thereby increasing capital and maintenance costs [5]. Finally, even the control of indoor space can be a rather complex process, including entry and exits of spectators and athletes, security stations, and rigid barriers. Given the number of logistics—space use, safety, and crowd control—that need specialized planning and resources, it is not surprising that expenses to put on indoor events grow [6]. Cumulative costs that account for the monetary burden of hosting indoor events demand strategic planning and cost management strategies. Costs associated with venue rental, maintenance, and operational logistics can pose significant challenges for organizers navigating budget constraints and seeking to maximize return on investment. It is critical that they do not compromise environmental safety due to cost constraints.

In exploring the difficulties associated with indoor sporting events, this chapter delves into the multifaceted nature of these challenges, drawing on empirical evidence and practical insights to illuminate key considerations for stakeholders across the sports industry. By examining these complexities through a comprehensive lens, we aim to foster a deeper understanding of the intricacies involved in delivering successful and safe indoor sporting events in a safe environment.

11.2 Case Anecdote/Challenge

Let's take Sarah as an example. She was always outstanding on the outdoor courts in her neighborhood, where she honed her skills under the open sky. But when she transitioned to indoor basketball at the collegiate level, she faced a whole new set of challenges.

The indoor court was smaller and more enclosed compared to the vast outdoor spaces she was used to. During her first game on the collegiate team, Sarah quickly realized that the dynamics of the game were different. The limited playing area meant she had to control her pace if she was running toward the baseline or sideline. If she ran to far out of bounds, she would run into the stands rather than surrounding open space.

In a crucial match against a rival team early in the season, Sarah found herself distracted by the overhead lights and the big screen scoreboard for advertisements. With defenders closing in, she looked up to receive a lob pass which was hidden in the overhead spotlights. She lost the ball, nearly lost her balance, and the team lost the game by a basket. Luckily, she learned from her mistakes and improved over time; however, this incident highlighted the unique challenges of indoor basketball.

Throughout the season, Sarah and her teammates worked tirelessly to adapt their strategies to the indoor environment. They focused on precise passing, quick transitions, and strategic positioning to make the most of the smaller court while avoiding turnovers and defensive mistakes as well as being aware of their environment. The artificial lighting and the noise from the enthusiastic crowd added to the intensity, increasing the pressure and requiring her to stay focused amid distractions. After a grueling regular season, Sarah and her teammates made it to the conference playoffs. Their first playoff game was scheduled for a rainy Saturday afternoon. Both teams had been preparing for weeks for this moment, hoping to move forward in their bracket and eventually reach the national championship. In an outdoor format, the game would have already been cancelled. In this case, just two hours before tip-off, the court started to accumulate water due to a leak in the roof. The sports directors from both teams got together and agreed that the game couldn't be played due to the unsafe footing until the roof was fixed. It was devastating for both teams, who had mentally and physically prepared themselves for this game, only to have it postponed due to the indoor environmental hazard. This not only burdened the athletes but also the families who had taken time out of their schedules to attend the game.

As Sarah looked back on her journey from outdoor courts to indoor arenas, she realized that mastering indoor basketball required a different mindset and a different set of skills that included being aware of critical environmental factors including lighting, footing, acoustics, and physical structure. It wasn't just about physical ability; it was about being tactically aware, working as a team, and being resilient in a controlled and high-stakes environment.

11.3 Epidemiology

In the world of human recreation, indoor sports have come a long way. It's a tale of necessity, innovation, and cultural transformation. Back in the day, people needed a way to enjoy physical activities without being at the mercy of the weather. That's when indoor sports stepped in, providing sheltered spaces where athleticism could thrive, no matter what Mother Nature had in store. The early chapters of this story show how indoor sports were closely linked to the development of suitable facilities. Gymnasiums and enclosed courts became havens where sports like basketball and volleyball found their groove. Not only did these spaces foster athletic competition, but they also became social hubs where communities came together.

Across different parts of the world, indoor sports unfolded with their own unique twists. In colder regions, indoor arenas offered a break from harsh winters, which provided year-round opportunities regardless of weather conditions. It introduced indoor sports such as ice hockey to regions that had never seen snow or outdoor ice. This accessibility fosters consistent training and competition as well as enhancing skill development and performance across various sports disciplines [7]. Indoor venues can also offer ideally controlled environments that minimize external variables, promoting fair competition and reliable performance metrics [8]. These advantages have been particularly beneficial in professional sports, where the reliability of scheduling and broadcast opportunities have boosted commercial viability and global reach [9].

Technological advancements played a big role in pushing this story forward. Improvements in flooring, lighting, and equipment have made indoor sports safer and optimized performance. New sports have emerged and evolved via the advent of indoor sporting venues, expanding beyond traditional outdoor disciplines. For instance, indoor volleyball which transformed from its beach origins in the late nineteenth century has become a popular international sport. Worldwide growth of volleyball was driven by its controlled setting that allowed for uniform play as well as competitive leagues and tournaments [10]. The introduction of purpose-built indoor facilities has led to transformations of sports such as indoor soccer (futsal), indoor football, ice hockey, and basketball. This, in turn, has led to faster-paced games with active fans' involvement [8]. The implication is that these innovations based on consistent and controlled environments can lead not only to the advancement of existing sports but can also create new formats of competition leading to new sports and variations that unique benefit by an indoor environment venue.

Ultimately, the story of indoor sporting events is a testament to human creativity and the unyielding pursuit of physical excellence. It shows how places once built for protection from the elements have become melting pots where athleticism, innovation, and community come together to celebrate the human spirit.

11.4 Unique Injuries and Challenges

Indoor sports take place in specialized facilities with hardwood floors or synthetic courts. These surfaces are great for precision and agility, but they also come with risks. The firm and unyielding nature of hardwood flooring, while it does provide excellent traction for quick movements and agility, can also contribute to higher risks for musculoskeletal injuries. Since many of the competition floors are temporary as the venue resets itself for different types of competition, the installation and undersurface cushioning can also play a role regarding injury risk. Research has consistently shown that the most common injuries are ankle sprains, knee ligament injuries, and muscle strains for athletes competing on these types of surfaces [11, 12]. Biomechanically, one of the main reasons for these injuries is related to the high friction coefficient of hardwood, which can lead to abrupt stops and changes in direction, placing increased stress and strain on joints and soft tissues. To review, friction is the force that resists a relative movement between two objects that are in direct contact with each other (Fig. 11.1). Therefore, the more friction between two objects, the greater the force that will be required to move one object over the other. A sticky, high friction floor allows for quicker changes in direction but also increases the loads across the joints of the lower extremity (Fig. 11.2).

More specifically, the repetitive impact of running, jumping, and landing on a hardwood surface or rigid surface can predispose these athletes to overuse injuries, including stress fractures and tendinopathies in the lower extremities [13]. In contrast, outdoor sports often take place in wide open spaces with grassy fields or dirt terrains that provide more inherent shock absorption. While the weather and surface texture and stability can be more unpredictable, the variation of outdoor surfaces may be more forgiving regarding impact loading the controlled environment of indoor venues. The softer surfaces cushion impacts and allow for more fluid movement which is a big contrast to the rigid indoor courts.

Head injuries and concussions are of significant concern in indoor sports due to the confined and often unforgiving environments of indoor venues. Recent research indicates that indoor sports may experience higher concussion rates than outdoor sports, which contradicts common expectations. For instance, sports like basketball, volleyball, and ice hockey exhibit significantly high concussion incidences. A notable study by Comstock et al. (2015) revealed that girls' indoor soccer had nearly double the concussion rate of outdoor soccer. Several factors could explain this trend: harder playing surfaces, the close proximity to walls and other players, and more frequent collisions in confined spaces [14]. Additionally, a report from the National Athletic Trainers' Association (NATA) highlighted that concussions comprised a higher percentage of total injuries in indoor sports like cheerleading (31.1%) and gymnastics (13.2%) compared to outdoor sports like football (7.4%) [15, 16]. Although further research is necessary, these findings underscore the need for heightened concussion awareness and prevention strategies in indoor sports settings. The close proximity of walls and other players in indoor settings can amplify the impact of collisions, falls, and contact incidents, leading to a higher incidence of head injuries. Additionally, the hard surfaces typical of indoor sports facilities, such as hardwood courts or artificial turf, offer little cushioning to absorb or dissipate impact forces, exacerbating the severity of head impacts.

Unlike outdoor sports, indoor sporting events face a unique challenge with ventilation, a stark contrast to outdoor sports where natural air circulation is readily available. Indoor venues must depend on mechanical systems to maintain optimal air quality and temperature, which are essential for the comfort and performance of athletes.

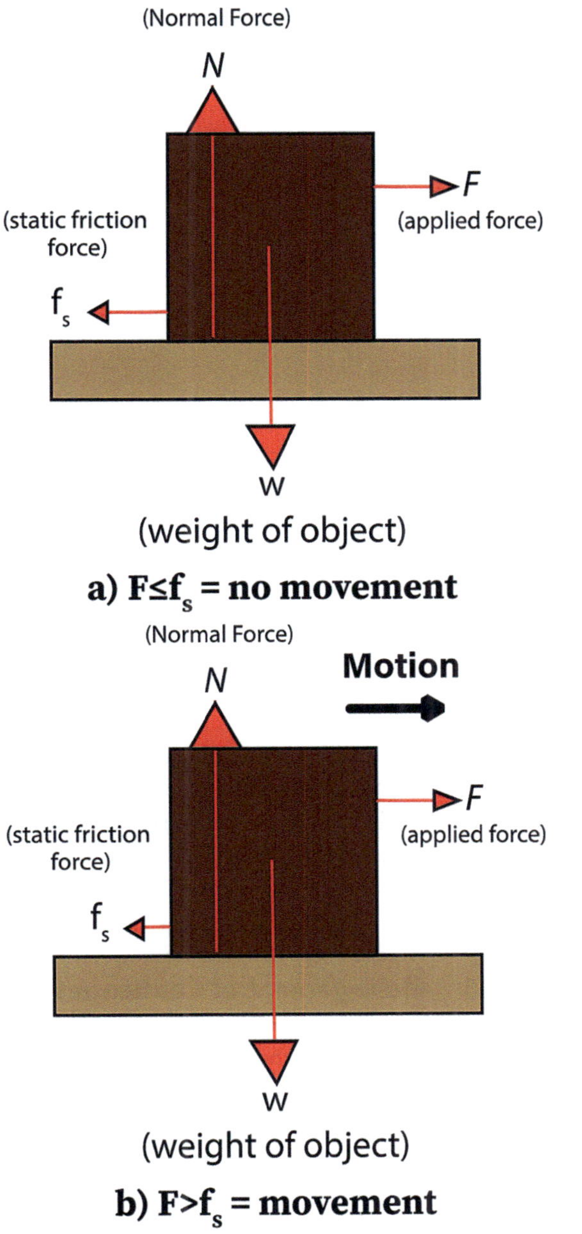

$$f_{(static)} \leq \mu_s N$$

$$f_{(kinetic)} = \mu_k N$$

Fig. 11.1 Friction force equation

Inadequate ventilation can lead to increased levels of carbon dioxide (CO_2), airborne pollutants, and humidity which can have a negative impact on both athletes and spectators. Studies have shown that poor ventilation in indoor arenas can worsen respiratory symptoms and compromise cardiovascular health in athletes, potentially hindering their endurance and recovery [17]. Additionally, the confined spaces and potential for overcrowding during events can place extra stress on ventilation systems, raising the risk of heat stress and discomfort for everyone present [18].

Material vs Rubber	Average Coefficient of Friction
Rubber	1.16
Dry Asphalt	0.90
Carpet	0.80
Wood	0.725
Dry Concrete	0.725
Wet Concrete	0.60
Wet Asphalt	0.50
Dry Grass	0.40
Wet Grass	0.30
Ice	0.15

Fig. 11.2 Coefficients of friction of different materials vs rubber

Infectious disease transmission is another distinctive challenge that we encounter with indoors events compared to outdoor events, primarily due to the enclosed, reduced air circulation and often more densely populated nature of indoor venues. The close proximity of athletes, coaches, and spectators heightens the risk of respiratory droplet transmission of pathogens like influenza viruses, adenoviruses, and coronaviruses [19]. Sharing equipment and facilities further increases the risk, as surfaces can serve as reservoirs for bacteria and viruses, enabling indirect transmission through fomites. Studies indicate that indoor environments with poor ventilation and air circulation can prolong the viability of airborne pathogens and lead to higher concentrations of airborne particles, exacerbating the risk of disease spread [20]. Moreover, the physical exertion associated with indoor sports can elevate respiratory rate and depth, potentially enhancing the spread of infectious droplets among participants [21]. The complexity of mitigating infectious disease risks in indoor sports is further compounded by factors such as international travel for competitions, which can introduce new pathogens and challenge existing infection control measures.

Apart from the physical challenges, indoor sports also take a toll on athletes' mental well-being, due to the controlled, often confined environments they compete in. Studies reveal that the limited space of indoor venues can increase mental pressure by intensifying performance scrutiny, accentuating mistakes, and amplifying the psychological effects of crowd behavior [22]. Additionally, the consistent lighting and predictable surfaces typical of indoor sports require greater technical precision and strategic focus, leading to increased mental fatigue and stress over extended periods [23]. The proximity of spectators and the acoustics of indoor arenas can also impact athletes' mental states, heightening arousal levels and affecting concentration and decision-making [24]. In contrast, outdoor sports provide athletes with more spatial freedom, exposure to natural elements, and variable conditions, which can reduce psychological stress and enhance adaptability and resilience. Recognizing these differing mental challenges is crucial for improving athlete preparation, performance strategies, and mental health support in both indoor and outdoor sports.

11.5 Management of Challenges

In order to navigate these differences, athletes, coaches, and organizers of indoor sports events need to take a holistic approach. This means focusing on injury prevention strategies, making sure the equipment and physical structures are optimized, maintaining good hygiene practices of the venue, and prioritizing the mental well-being of everyone involved. By doing so, they can create a safe and fulfilling athletic experience in the world of indoor sports. In the world of indoor sports, where competition echoes off polished floors and the stakes are high in tight spaces, managing the unique challenges requires a careful blend of strategy, care, and foresight.

Picture yourself stepping into a buzzing indoor arena, where the shiny hardwood floor glows under bright lights and athletes gear up for intense physical activities. The first line of defense lies in the very foundation beneath their feet. Regular inspections and meticulous maintenance of the surfaces ensure that risks like slippery floors or uneven playing fields are quickly taken care of. It's a small but vital detail that sets the stage for safe athletic endeavors.

Looking up, a quick inspection of the lighting system may reveal individual or sections of lights that are out leading to shadows for the athletes to overcome on the field of play. Bright lights or flashes from cameras within the athletes' line of site can blind them during an important part of their competition. If recognized and adjustable, the lighting should be changed to decrease risk and the audience reminded to avoid flash photography.

Ventilation whispers through the rafters as a critical lifeline against stagnant air and respiratory discomfort. Improving airflow within these enclosed spaces becomes crucial, not just for comfort but also to protect against airborne pathogens that thrive in close quarters. If the ventilation system is adequate or goes down, opening windows or doors with addition of mobile fans may improve air quality and circulation to a degree. Temperature control also goes hand in hand with ventilation, creating an environment where athletes can perform at their best without overheating or feeling chilly drafts. If temperatures rise to an unsafe level or if air circulation is not functioning, particularly during a respiratory pandemic, the event may need to be postponed.

Down on the court or mat, another layer of preparation unfolds. Athletes put on specialized footwear designed to grip the smooth surfaces with precision. Coaches emphasize the importance of proper technique and form, knowing that the right movements can mean the difference between victory and injury. Warm-up rituals and meticulous cooldown routines become rites of passage, priming muscles and joints for the demands of rapid pivots, sudden stops, and explosive jumps that define indoor sports. The footwear should be adjusted to the specific environment to both optimize performance but assure athlete safety.

Best practice when managing an indoor sports event is to maintain a routine checklist for both home and away events. Inspect the lighting, physical plant, and flooring with special attention regarding injury risk of athletes. Perform a walk-through of the facility. Assess ventilation and temperature and plan to adjust or be prepared for at-risk athletes if the ventilation is substandard or temperatures are elevated. Create an emergency plan that includes where you will care for an injured athlete. How far away is the training room or locker room? Is it stocked with necessary medical equipment or will you need to carry your game bag? Have you considered catastrophic injury on the playing field? Where is the AED? Where will local emergency responders have access to the playing field in case you need assistance? Where is the egress/evacuation point and location of the ambulance if extrication and transport become necessary?

Many facilities will have an emergency plan in place that addresses spectators and athletes separately. Making yourself aware of the existing emergency plan will avoid excessive efforts of reinventing the wheel. With that said, modifications of the emergency plan may be necessary depending on the specific sport and event that you will be covering. Do you or your governing body require emergency transport on-site or on call? These arrangements will need to be made well in advance including a clear understanding of where the emergency vehicle will park, where the emergency responders will sit, when they should approach an injured athlete (usually the team's medical team is first responder on court), how the emergency responders will be signaled to join in the care of an injured athlete, who is in charge during any injured athlete movement or extrication, how the injured or immobilized athlete will transitioned to the emergency vehicle (what is the actual pathway), and who will accompany the athlete during transport.

11.6 Prevention of Injuries/Illnesses in Indoor Sports

Ultimately, attention to a variety of environmental details will serve as the best pathway to reduce the risk of injury and illness in indoor sporting venues. Being aware of the unique challenges faced by performing in indoor environments is the critical first step. Attention to the details of lighting, air quality and circulation, playing surface, noise, proximity of spectators, unique structural elements of the facility can identify at-risk areas that the sports medicine provider can either address or make the athlete and coaches aware of.

Documenting and recognizing the actual incidence of certain illness and injury patterns in the indoor environment not only increases awareness of the problem but also establishes a baseline for injury prevention and illness reduction. Such recognition opens the door to intervention that can reduce injury risk and improve athlete safety. Optimizing lighting, assuring that the flooring is well cushioned and consistent in the field of play, padding or moving physical structures that place the athlete at risk, and optimizing environmental controls of air quality and temperature can reduce injury and illness. Finally, an assessment of actual impact of intervention is essential to confirm a positive outcome and guide the next steps of optimizing the impact of injury prevention in the indoor environment. Consistently applying the injury prevention model originally proposed by van Mechelen allows the healthcare team to be silent guardians of athlete well-being, providing not only treatment but also guidance on prevention and rehabilitation to keep athletes in the game (Fig. 11.3).

Continued education should run through the veins of every team member and staff. Safety seminars and workshops become a rite of passage, reinforcing the importance of injury prevention and health maintenance. Policies—clear, concise, and unwavering—are integrated into the fabric of daily routines, ensuring that everyone, from the youngest athlete to the seasoned coach, understands their role in maintaining a safe and respectful environment. Data-driven insights fuel ongoing assessments, revealing trends and near misses that guide continuous improvement. It's a journey of evolution, where each practice session and every game become a stepping stone toward refining strategies and fortifying defenses. This is true for indoor sports but all environments. In this story of indoor sports management, every detail—from the gleam of the floor to the warmth of a handshake—is a testament to the dedication and vigilance required to ensure safety, foster excellence, and uphold the spirit of competition within the revered confines of the indoor arena.

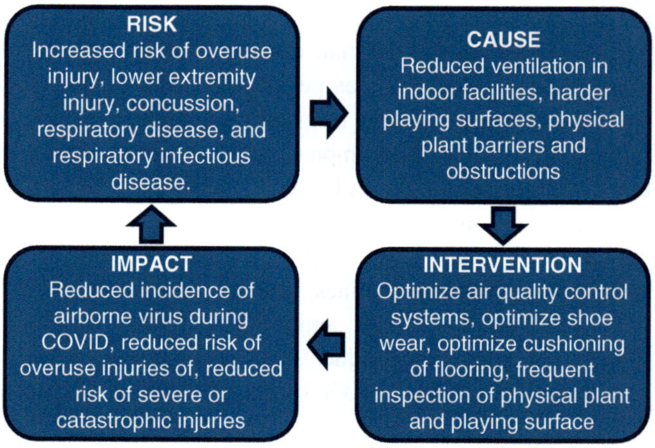

Fig. 11.3 Assessment of indoor sport using van Mechelen model

11.7 Take-Home Pearls

Indoor sports come with their own set of hurdles. Athletes have to navigate on polished surfaces where every twist, jump, and pivot carries the risk of impact-related injuries like fractures, sprains, or collisions, especially since they're in such close proximity to their competitors.

Indoor sports appear to have elevated risks of overuse injuries, concussions, respiratory related diseases, certain infectious disease, and mental health stress compared to outdoor sports.

While the controlled climate protects them from the elements, it also requires careful attention to lighting, ventilation, and temperature management to ensure the athletes stay comfortable and healthy.

Managing the facilities is crucial to ensuring safety in indoor environments. Regular maintenance keeps the playing surfaces hazard-free, improved ventilation systems tackle the challenge of stagnant air that can cause respiratory issues, and physical plant structures near the field of play need to be padded and removed. These measures not only protect the athletes but also create an environment that fosters peak performance.

Indoor facilities introduce unique benefits and challenges for the healthcare team including the potential availability of controlled locker rooms, training rooms, and medical facilities but the challenge of how physical structures and access may hazard emergency response, extrication, and evacuation of injured athletes.

Looking ahead, the future of indoor sports is filled with promise and innovation. Advances in facility design and technology offer the potential for safer and more efficient venues that enhance both performance and the spectator experience. The integration of data analytics could revolutionize injury prevention strategies by providing insights that optimize training and enhance player longevity.

References

1. Mendes FA, Grizzo LT, Lopes AJ, Jardim JR, Nascimento OA. Influence of indoor environmental quality on respiratory health in primary schools. Environ Monit Assess. 2019;191(10):Article 624. https://doi.org/10.1007/s10661-019-7765-1.
2. Chung KF, Rondo DM. Indoor air quality and its effects on respiratory health among children and athletes. Paediatr Respir Rev. 2016;18:66–71. https://doi.org/10.1016/j.prrv.2015.12.004.
3. Burge S. Sick building syndrome. Occup Med. 2010;60(3):205–6. https://doi.org/10.1093/occmed/kqq038.
4. Serafini A, et al. Energy consumption analysis of indoor sporting arenas compared to outdoor stadiums. J Sustain Energy. 2019;8(3):112–26. https://doi.org/10.1080/14786451.2019.1577330.
5. Lamont M. The anatomy of sociocultural costs: on the organization of conventions, conferences, and other indoor events. Annu Rev Sociol. 2017;43:159–79. https://doi.org/10.1146/annurev-soc-060116-053310.
6. Horne J, Manzenreiter W. An introduction to the sociology of sports mega-events. Sociol Rev. 2006;54(2):1–24. https://doi.org/10.1111/j.1467-954X.2006.00659.
7. Gerdin G, et al. Impact of indoor sports facilities on physical activity, health and social connectedness: a systematic review. Sports Med. 2019;49(11):1765–82. https://doi.org/10.1007/s40279-019-01147-1.
8. Hoppeler H, Baumgartner L. Skeletal muscle plasticity in response to selected environmental and physiological stimuli. J Appl Physiol. 2016;90(3):1159–66. https://doi.org/10.1152/japplphysiol.01195.2000.
9. Lamont M. The impact of indoor sports venues on the global sports industry: a critical review. J Sport Manag. 2017;31(4):459–76. https://doi.org/10.1123/jsm.2016-0261.
10. Horne J, Manzenreiter W. An introduction to the sociology of sports mega-events. Sociol Rev. 2006;54(2):1–24. https://doi.org/10.1111/j.1467-954X.2006.00659.x.
11. Alonso JM, et al. The implementation of an injury surveillance system in professional soccer: recording multiple injuries and days lost. Am J Sports Med. 2019;29(2):140–6.
12. Gabbe BJ, et al. Predictors of lower extremity injuries at the community level of Australian football. Clin J Sport Med. 2016;17(2):74–80.
13. Videbæk S, et al. The effect of eccentric hamstring strength training on sprint performance and injury risk factors in competitive female soccer players. Am J Sports Med. 2015;31(5):693–9.

14. Comstock RD, Currie DW, Pierpoint LA, Grubenhoff JA, Fields SK. An evidence-based discussion of heading the ball and concussions in high school soccer. JAMA Pediatr. 2015;169(9):830–7.
15. National Athletic Trainers' Association. Athletic Training Services: An Overview of Skills and Services Performed by Certified Athletic Trainers; 2019
16. McCrory P, et al. Consensus statement on concussion in sport—the 5th international conference on concussion in Sport held in Berlin, October 2016. Br J Sports Med. 2017;51(11):838–47.
17. Garrigues E, et al. Effect of indoor air quality in homes on respiratory health: a systematic review. Int J Environ Res Public Health. 2018;15(7):Article 1386. https://doi.org/10.3390/ijerph15071386.
18. Hou H, et al. Impact of ventilation systems and confinement on indoor air quality and health in homes. Build Environ. 2016;106:37–48. https://doi.org/10.1016/j.buildenv.2016.04.024.
19. Hui DS, et al. The continuing 2019-nCoV epidemic threat of novel coronaviruses to global health—the latest 2019 novel coronavirus outbreak in Wuhan, China. Int J Infect Dis. 2018;91:264–6. https://doi.org/10.1016/j.ijid.2020.01.009.
20. Sundell J, et al. Ventilation rates and health: multidisciplinary review of the scientific literature. Indoor Air. 2011;21(3):191–204. https://doi.org/10.1111/j.1600-0668.2010.00703.x.
21. Morawska L, Cao J. Airborne transmission of SARS-CoV-2: the world should face the reality. Environ Int. 2020;139:Article 105730. https://doi.org/10.1016/j.envint.2020.105730.
22. Schinke RJ, et al. Examining the role of mental skills in the link between sport-related concussion symptoms and psychological well-being. J Clin Sport Psychol. 2020;14(2):245–64. https://doi.org/10.1123/jcsp.2019-0019.
23. Swann C, et al. The influence of environmental factors on mental health and performance in athletes. Sports Med. 2017;47(9):1687–700. https://doi.org/10.1007/s40279-017-0696-7.
24. Lidor R, Ziv G. Physical and psychological effects of indoor versus outdoor environments for exercise. Int J Sport Exerc Psychol. 2016;14(4):375–88. https://doi.org/10.1080/1612197X.2014.971194.

It's Polluted

12

James D. Oosten, Matthew A. Siegel, and Mark R. Hutchinson

12.1 Introduction

Sports are among the most popular activities in the world and are about as pervasive as the air we breathe. Unfortunately, pollution in the environment is equally common across the globe and poses a challenge to athletes everywhere. Pollution causes inflammation, particularly in the airway, through oxidative stress from a combination of increased reactive oxygen/nitrogen species (RONS) and decreased levels of antioxidants, often the result of chronic exposure [1]. Athletes are particularly impacted by air pollution as they breathe over 12 times as much air during exercise-induced ventilation compared to the resting state [2]. During exercise, open mouth breathing allows small particulates within the air to bypass the filters of the nose, and increased inspiratory force deposits pollutants deeper into the respiratory tract, at a 4.5 times higher rate than during rest [3]. This has been associated with an increase in exercise-induced bronchoconstriction (EIB), cardiovascular disease, emergency department visits for respiratory symptoms, and new or worsening bronchodilator requirements for individuals with or without preexisting asthma, among other conditions [4–7]. Figure 12.1 provides an overview of important considerations of air pollution and physiology during exercise [8].

Aquatic athletes are susceptible to an additional subset of local pollutants, microbes, and chemicals that may affect the skin, respiratory, or gastrointestinal system during or after competition. These are ingested directly, inhaled as volatile solutes, or absorbed through the skin. Studies estimate ingestion during a training session can be as high as 100 milliliters and on average is ~16 mL for adults and 37 mL for children [9]. Incidence of illness from contaminated waters correlates positively with longer time spent in the water and intensity of swimming, such as seen in competitive swim meets [10].

The public discussion regarding pollution in sports medicine gained significant traction in anticipation of the 1984 Olympics in Los Angeles, California, given the concern regarding smog [11]. Medical experts recommended avoiding pollution and administering oxygen and vitamins C and E for therapeutic intervention, which have since been validated as possible strategies to manage pollution-exacerbated respiratory disease [12]. Since then, progress has been made on

J. D. Oosten
Orthopaedic Surgery PGY-2, University of Illinois, Chicago, IL, USA

M. A. Siegel
Orthopaedic Surgery, PGY-4, University of Illinois at Chicago, Chicago, IL, USA

M. R. Hutchinson (✉)
UIC Sports Medicine Center, Department of Orthopaedics, University of Illinois at Chicago, Chicago, IL, USA
e-mail: mhutch@uic.edu

M. R. Hutchinson, S. Rocha Piedade (eds.), *Sports Medicine: It's Environmental*,
https://doi.org/10.1007/978-3-031-84493-5_12

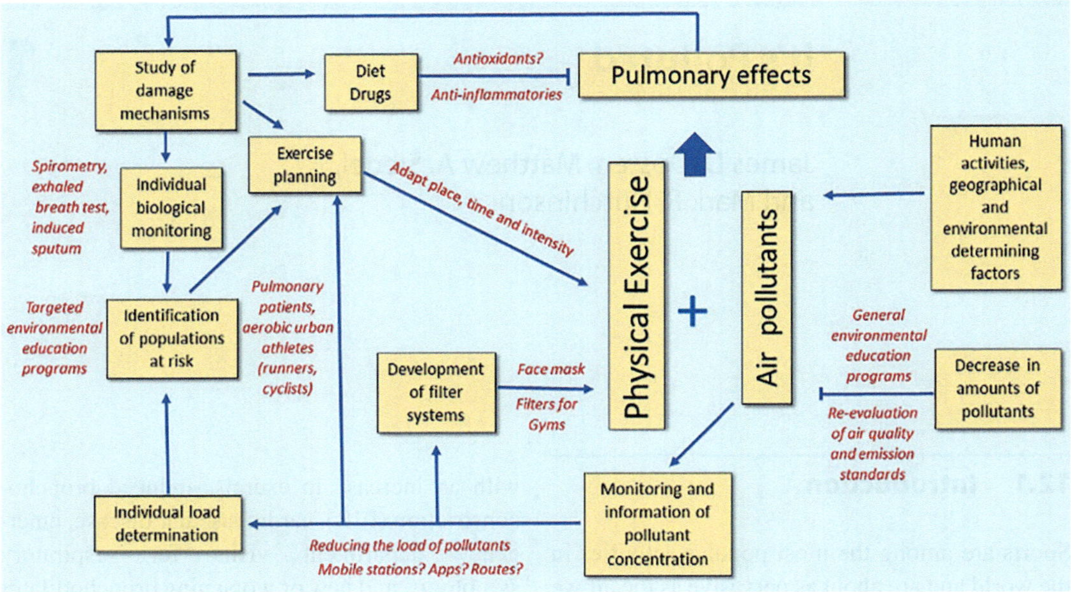

Fig. 12.1 Overview of future goals and strategies to optimize the performance of exercise in polluted conditions. Reproduced with permission from Araneda et al. [8], under the terms and conditions of the Creative Commons Attribution (CC BY) license (http://creativecommons.org/licenses/by/4.0/)

the local, regional, and international level to identify education, treatment, and prevention strategies to mitigate the impact of pollution on athlete health and performance at all levels. This was critical to facilitating the 2008 Summer Olympics in Beijing amid the international concerns regarding China's air quality from industry and overpopulation. Similar principles were applied to the Rio 2016 Olympic Games, instead of focusing on contamination of the water with very high viral and bacterial loads. Controversy surrounding the Paris 2024 Olympic Games arose during pre-Olympic swimming competitions in 2023, which were cancelled due to pollution concerns within the Seine. The river was intended to be a focal attraction of the French Olympic planning committee, but the World Aquatics organization was unable to declare the water quality fit to compete in, requiring allocation of more resources toward preparing the river for the global event.

12.2 Case Anecdote: An International Challenge

A 23-year-old distance runner preparing for the Beijing Olympic Games has a known history of asthma which is usually well controlled with occasional use of an albuterol inhaler. The runner commonly trains in the open air of a mountain environment to take advantage of altitude training. They are concerned about the coming competition in the known air pollution of Beijing. They would like to know if they should be concerned. Should they train differently? Should they be on prophylactic medication? What should they do if they have an event in China to assure that they optimize performance?

Despite global efforts to improve air quality, almost seven million deaths across all ages in 2019 were accredited to air pollution of particulate matter (PM) 2.5 micrometers or less [13]. Public concern regarding air pollution escalated

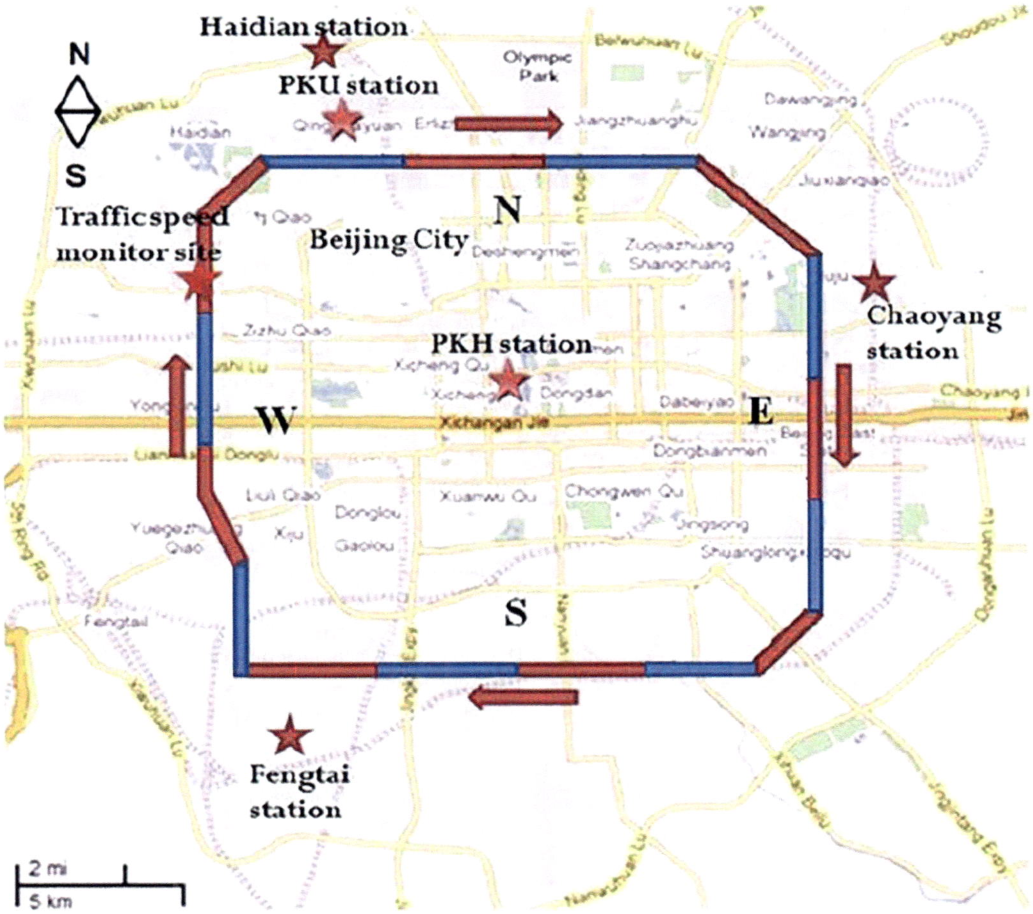

Fig. 12.2 Map of the City of Beijing. The 4th Ring Road was chosen as the sampling route. For analysis purposes, this route was binned into 16 sections of equal length seen here colored red and blue. The red stars show meteorological stations and places where traffic speed was recorded (version: Google map 2009). (Reproduced from Wang et al. [15], under the Creative Commons Attribution 3.0 License)

in the months leading up to the 2008 Olympic Games in Beijing, China, which has some of the highest concentrations of PM2.5 in the world, mostly stemming from motor vehicle emissions [14].

The Chinese Government and the Beijing Municipal Environmental Protection Bureau (BJEPB) took calculated precautions to ensure safety for its athletes and spectators during the Olympic games (August 8–24, 2008) and the Paralympics (September 6–17, 2008). Starting on July 20, 2008, Ring road traffic was regulated via license plate number—only even-numbered plates were permitted to drive on even calendar days and vice versa with odd-numbered plates (Fig. 12.2) [15, 16]. Regulations also limited automobile entry into Beijing to cars with Euro 2 emissions standard, while diesel and other significant polluting cars were forbidden (Table 12.1). Additional metro and bus routes were created for the public, and the "Olympic Lanes" were created granting exclusive access to vehicles involved in the Games. Moreover, many nearby factories and construction sites ceased and/or reduced production or were relocated months in advance of the Olympic games [15, 16].

Table 12.1 Sampling and starting dates of different air pollution control measures before, during and post 2008 Olympics and Paralympics (compiled from Beijing EPB released information on web page, The 14th control strategy on Beijing air quality, 2008)

Periods	Detailed periods[a]	Starting dates[b]	Control measures[c]	Sampling dates[d]
Before full scale control (Pre-20 Jul 2008)	Before full scale control (Pre-20 Jul 2008)	1 Mar 2008	Introduce new vehicular emission standard, equivalent to Euro 4	18 Jul 2008
				19 Jul 2008
		Not defined[e]	Relocating heavy industrial polluters, etc. Shou gang steel factory and other factories in south area of Beijing; install desulfurization facility in factories around Beijing	
			Implementing low fugitive emission facilities at more than 1000 gas stations	
		23 Jun 2008	50% of government cars were not allowed to drive	
			Diesel and heavy polluting vehicles not allowed to drive in Beijing	
			Only those vehicles meeting emission standards equivalent to Euro 2 were allowed to enter Beijing	
Full scale control (20 Jul 2008–19 Sep 2008)	Full scale control: before Olympics (20 Jul 2008–7 Aug 2008)	20 Jul 2008	Starting the full scale control	18 Jul 2008
			The odd/even plate number rule for traffic control	22 Jul 2008
			Stricter control on vehicles entering in Beijing	28 Jul 2008
			Reduce or stop production at certain factories surrounding Beijing	30 Jul 2008
				4 Aug 2008
	Full scale control: during Olympics (8–23 Aug 2008)	8 Aug 2008	Extra 20% of governmental cars were not allowed to drive	12 Aug 2008
			Stop outdoor construction activities	14 Aug 2008
			Temporally close some gas stations	16 Aug 2008
			Increase bus fleet and transit frequency	18 Aug 2008
				22 Aug 2008
	Full scale control: Between Olympics and Paralympics (24 Aug 2008–6 Sep 2008)		The same as Full scale control: before Olympics	25 Aug 2008
				27 Aug 2008
				28 Aug 2008
				29 Aug 2008
				4 Sep 2008
				5 Sep 2008
	Full scale control: during Paralympics (7–19 Sep 2008)		The same as Full scale control: during Olympics	8 Sep 2008
				9 Sep 2008
				10 Sep 2008
				11 Sep 2008
				18 Sep 2008
				19 Sep 2008

(continued)

Table 12.1 (continued)

Periods	Detailed periods[a]	Starting dates[b]	Control measures[c]	Sampling dates[d]
Post full scale control (After 20 Sep 2008)	Post full scale control (After-20 Sep 2008)	20 Sep 2008	Lifting of regulations adopted from 20 July	22 Sep 2008
			Control 20% of private cars based on the last digital of plate number	25 Sep 2008
				26 Sep 2008
				28 Sep 2008
				30 Sep 2008
				2 Oct 2008
				6 Oct 2008

[a]Divided the full scale control period into four periods: full scale control before Olympics, full scale control during Olympics, full scale control between Olympics and Paralympics and full scale control during Paralympics. Therefore, six periods were formed based on different control measures. Details please see Sect. 2.5

[b]Showed the starting dates of different control measures on the air pollutants in Beijing and were categorized to the six detailed periods

[c]Showed the chief contents of control measures on air pollutants made by Beijing EPB corresponding to different periods

[d]Listed the sampling dates with 31 days in total, covering the six detailed periods

[e]Means the starting dates were unclear but all of these control measures should be finished before 31 June, 2008

When compared to values prior to July 20, 2008, significant decreases were seen during the Beijing Olympic Games in pollutant concentrations, including 54%, 41%, and 70% reductions in carbon monoxide (CO), nitric oxide (NOx), and sulfur dioxide (SO_2). Benzene, toluene, ethylbenzene, and xylene (BTEX) were reduced by 66%, black carbon (BC) by 12%, and PM1.0 surface area (SPM1) by 18% [15]. After restrictions were lifted on September 20, 2008, these values trended back toward their baselines.

The quasi-experimental nature of the Beijing Olympics raises the question of ideal training and competition environments. Outpatient visits for asthma decreased to approximately 7.3 per day from 12.5 per day during the Olympic Games (relative risk reduction 0.54) [6], inflammatory biomarkers in young individuals (mean age 24 years) were reduced [17], and emergency department visits for cardiovascular emergencies decreased along with PM2.5 trends, albeit with an approximately 7 day delay in effect [18].

In the context of the Beijing Olympic Games, data is lacking that clearly defines how improvement in air quality impacted Olympian performances. Given the current evidence that shows how it alters physiology, it can be assumed that, had extensive measures not been taken to reduce pollution that summer, air pollution may have played a larger role in athlete performance during the 2008 Games. Future event organizers should be familiar with the evaluation of air quality and preventative measures to optimize the health of both their athletes and their spectators during large-scale events.

12.3 Epidemiology and Environmental Pollutants

Pollutants often vary based on rural, suburban, or urban settings, traffic and industrial emissions, and regional ecosystems. The common pollutants encountered are discussed below, with WHO air quality guidelines summarized in Table 12.2 [19].

Particulate matter (PM) is known to cause injuries to the airway and vasculature, impair mucociliary clearance, and can impact respiratory development in children [20, 21]. Particulate matter is classified based on diameter, as the toxicity of various PM depends upon the surface area, exposure rate, and charge of the particles [22]. Coarse particles are defined as 2.5–10 micrometers in diameter (PM10, commonly from industrial and steel mill exhaust), fine particles 0.1–2.5 micrometers (PM2.5, commonly from

Table 12.2 Air pollutant concentration values based on World Health Organization's air quality guidelines. These are designed to offer guidance in reducing the health impacts of air pollution based on expert evaluations of current scientific evidence

Pollutant	Time frame	Guideline mean value
PM2.5	Annual	5 μg/m³
	24 h	15 μg/m³
PM10	Annual	15 μg/m³
	24 h	45 μg/m³
O3	8 h	100 μg/m³
	Peak season	60 μg/m³
NO_2	Annual	10 μg/m³
	24 h	25 μg/m³
	1 h	200 μg/m³
SO_2	10 min	500 μg/m³
	24 h	40 μg/m³
CO	1 h	30 mg/m³
	8 h	10 mg/m³
	24 h	4 mg/m³

Note that there are other types of PM, including PM1 and ultrafine particles (UFP), black/elemental carbon (BC/EC), and sand and dust storms (SDS) that do not yet have sufficient data for recommendations to be made. Data adapted from World Health Organization Global air Quality Guidelines, 2021 World Health Organization, with modifications

furnaces, smelters, and motors) and ultrafine particles <0.1 micrometers (PM.1, commonly from diesel engines) [23, 24]. Particles greater than 10 micrometers are typically filtered within the nasopharynx and are thus less toxic to pulmonary function. PM 2.5 and smaller may penetrate deeply into the interstitium of the lung and accumulate over time [22]. Particles may even penetrate the circulatory system given the epithelial compromise from high stress within an actively ventilating lung, leading to systemic oxidative stress. PM exposure at high levels for intermediate periods of time (20 min +6 min maximal work test or 30 min) may reduce vasodilation by 8% and slow upper extremity muscle reoxygenation by 55%, respectively, affecting acute performance and recovery capabilities [24–26]. Ultrafine particles are thought to be the most toxic, which is problematic for athletes exposed to traffic-related air pollution (TRAP) while training or competing near major roadways [27]. There is also a high prevalence of PM found in indoor ice rinks from Zamboni and space heater usage [28] and in ski lodges and shops from waxing equipment [24].

Nitrogen dioxide (NO_2) is a precursor to particulate matter and is common in vehicular pollution [1]. It has been found to decrease a subset of lymphocytes after exposure during light exercise greater than 15 min, possibly contributing to athlete immunosuppression. Urban areas typically range 20–90 μ/m3 (0.01–0.05 ppm) but can reach up to 10 times that near busy roads.

Sulfur dioxide (SO_2) is a gas that has minimal effects on pulmonary function at ambient levels less than 2 parts per million (ppm). SO_2 may have synergistic effects on the bodies response to pollutants such as ozone (O_3), especially in subjects with previous pulmonary pathologies [1]. Asthmatics are susceptible at around 0.5 PPM. SO_2 concentrations are similar among urban and rural communities but vary based on geographic locations when comparing different countries [19].

Ozone (O_3) is formed by a series of complex reactions that take place in the atmosphere after exposure of NO_2 and volatile organic compounds to solar radiation [19]. Ozone is highly concentrated in large cities, particularly in America and Europe. Studies have found ozone levels to negatively correlate with levels of NO_2 and CO_2 and positively with sulfur and PM 2.5 [17]. Ozone exposure in hot environments has been shown to increase airway inflammation, which is more prominent during higher intensity and longer periods of exercise [8, 29]. Ozone has been shown to increase dyspnea and chest tightness and decrease lung function parameters during exercise in both trained and untrained subjects [8].There are no apparent synergistic effects of NO_2 and O_3 or PM on airway function [1].

Carbon monoxide is a pollutant with a high diffusion capacity that can bind to hemoglobin and cause a relative hypoxemia by limiting oxygen delivery to muscle tissue [8]. This has been shown to decrease performance in healthy subjects by limiting energy production and physical effort at maximum intensity exercise, although this effect is not present during submaximal intensity exercise [30].

The other major styles of pollutant encountered by athletes are those within aquatic environments.

Trichloramines (NC13) are the most common disinfection by-products (DBP) and are volatile aerosolized chemicals that present at the surface after a reaction between chlorine disinfectants and nitrogenous compounds, such as those found in sweat and urine, soap, or cosmetic residues. They can cause allergic hypersensitivity responses when inhaled acutely and structural epithelial changes with chronic exposure [2]. Experts suggest NC13 concentrations should remain under 0.5 mg/m^3, a number which is often exceeded in the air at indoor swimming pools despite appropriate water concentrations [1].

Several bacterial, viral, and fungal contaminants can be found in pools and open water sources. They may arise from shedding of fecal or nonfecal matter into the waters and often persist despite disinfection efforts. Disease outbreak can occur in water bodies with sewage overflow or flash flooding, especially after heavy rainfall [31]. Enterococcal concentrations should be below 200 per 100 mL of water and *E.coli* below 500 per 100 mL [31]. Coastal water and rivers are more likely to contain viral pathogens [32]. A breakdown of specific species, water treatment, and testing recommendations can be found in the WHO Guidelines for Safe Recreational Water [9, 33].

12.4 Injuries and Illnesses: Pollution's Impact on Athlete Performance

Numerous studies have used analysis of soccer, baseball, and football players to examine how pollution can impact sport-specific functions. Increases in air pollutant levels have been associated with decreases in pass or throw accuracy, scoring abilities, and overall task completion and decision-making speed [34–37]. Athletes can experience respiratory symptoms and decreases in pulmonary function while performing activities that range in intensity from walking to competitive cycling [8]. Short-term exposure (i.e., 20–30 min) to dense levels of air pollution (i.e., > 60,000 particles/cm^3 of high PM or 100–300 ppb O_3) can worsen healthy athlete respiratory function by 5–10% and overall performance by 3–5% [25]. This effect has been shown to last up to 3 days [21]. Even when within acceptable limits, changes in PM10 concentration were shown to decrease marathon performance in women by 1.4% for every 10ug/m [3] increase [3]. Cusick et al. found that elevated PM2.5 + O_3 combined levels, even in the context of "good-to-moderate" air quality index (AQI), were associated with slower collegiate male race times [38]. NO_2 has been shown to impair athlete response times, attention span, and cognitive flexibility [36]. Some studies suppose that exercising in presence of TRAP leads to a failed response of neurotrophic factors that are typically increased with exercise, thus hindering cognitive improvements in athletes [39]. Effects on performance can last for over 24 h post-exposure; coaches and players should be mindful of minimizing unnecessary exposure to public transportation, kitchens, bars, casinos, or bonfires immediately before major competitions [24]. When traveling or resting during times of poor AQI, vehicle and building windows should remain closed, air filters should be used, and masks should be worn [40].

The presence of any exercise-induced respiratory symptoms should not deter sports participation, unless specified by a specialist physician. Studies have shown that the effects of regular aerobic exercise over time do outweigh those of standard levels of pollution and still reduce air pollution-related mortality [29, 41]. In fact, continued sport participation is encouraged as it improves minute ventilation and reduces overall bronchoconstrictive triggers and response [42]. Some even propose that lung function may benefit from intermittent exercise in polluted air as the lungs become conditioned to oxidative stress [8]. Acclimation to ozone, but not other pollutants, has been shown to occur in as little as 4 days. O_3-induced performance deficits can improve with daily consecutive exposure to O_3, but this effect does not last, and risks and benefits of O_3 acclimation protocols prior to competition are not yet well understood [40]. Experts still recommend limiting or avoiding intense activity during periods of very poor air quality [40]. Presently, the use of targeted respiratory muscle training (RMT) therapies has not been consistently shown to improve athletic performance or decrease respiratory symptoms [43]. Regardless, worsening pollution encountered globally requires coaches, trainers, and physicians to modify existing training regimens and sideline preparedness strategies.

Current evidence that quantifies changes in swimmer performance based on chlorine-related airway hyperresponsiveness is lacking, mainly because most swimmers practice and compete in similar pool environments. Studies have shown that all pulmonary function test (PFT) values are significantly reduced in those acutely and chronically exposed to aerosolized chlorine [44]. Further, forced expiratory volume/forced expiratory capacity (FEV1/FVC) ratios in swimmers may decrease by 2–5% from their on-land baseline, depending on sex [45]. Pulmonary symptoms among swimmers are more prevalent during competition-speed intensity swimming. Chlorine levels should remain near the minimum acceptable disinfectant concentration to avoid hindering pulmonary function, and training should be conducted at sub-competition intensity levels to establish a gradual tolerance for longer training sessions.

12.5 Management of Pollution-Related Disease States

Elite athletes are widely believed to be somewhat immunosuppressed, which may contribute to their susceptibility to pollution-related illnesses [32]. The most relevant conditions to be familiar with are summarized in Table 12.3. Case-specific symptoms and risk factors should be accounted for as atypical presentations are common and an in-depth discussion on diagnosis and treatment is beyond the scope of this work. Extensive research has been done on exercise-induced bronchospasm (EIB) and airway hyperactivity, as this is one of the most common pollution-related disease states among high-level athletes. Table 12.4 provides a succinct summary on the current level of evidence for non-pharmacologic strategies for management of this common sequelae [2]. Presently, there is insufficient evidence to prove that EIB alone significantly worsens athletic performance beyond symptomatic discomfort [46].

Any athlete planning on competing in high-level competitions, particularly at the national or international levels, should have their medication list reviewed by a physician and screened against their regulatory organization's anti-doping laws [47]. Resources exist including the WADA Prohibited List and Global Drug Reference Online (Global DRO) website, where visitors, including sports medicine physicians, can search for specific information on medications that may or may not be prohibited. Commonly prescribed medications like pseudoephedrine and oral corticosteroids are banned, and others such as inhaled beta agonists may need documentation of medical necessity.

Table 12.3 Systems-based breakdown of common pollution-related illnesses, along with common triggers, presenting symptoms, and preventative and treatment measures for each

System affected	Illnesses	Trigger	Athletes at risk	Details	Diagnosis	Prevention Strategy	Acute management	References		
Pulmonary	Asthma, EIB, and AHR	Multiple air pollutants, including PM, SO_2, O_3, and metal particulates	Common in competitive athletes, particularly endurance	Dyspnea out of proportion, wheezing, chest tightness, cough (often transient, can be post-exercise)	Confirm presence of AHR via specialist consultation	Exposure avoidance most important	ICS for baseline, +/− IBA[a] similar to asthma management based on symptom severity/frequency, although ICS underutilized	Weisenberg [23], Betterfeld		Stellingwerff [52]
		Cold/dry air	Most common medical condition among Olympians (8–12%)	Atypical: diffuse stomachache, fatigue	Bronchodilator test, EBH test, Methacholine test, exercise test	Adequate warm-up (10–15 min), target HR 50–60% of maximum HR	SABA[a]: as needed. Regular use leads to tachyphylaxis, loss of bronchoprotective effects	Hough 1994		Ora
			20–50% of skaters		Differential diagnosis: panic attack, vocal cord dysnfunction, cardiac arrhythmia	Warmups: short-term, high and viriable-intensity warmups alleviate EIB severity	Cromolyn or antileukotriene. anticholinergics only for chronic cases			Mckenzie [47]

(continued)

Table 12.3 (continued)

System affected	Illnesses	Trigger	Athletes at risk	Details	Diagnosis	Prevention Strategy	Acute management	References		
		NCl_3	76% of swimmers			Non-chlorine disinfection of pools, ie using copper, silver or Ozone INC monitoring of NCl_3 levels		Harder [31]	Chamberlain [32]	Russo
Eyes, ear, nose, throat	URI	Inflammation from irritants, i.e., NCl_3	Competitive athletes (4.5 higher incidence than recreational), vitamin D deficiencies	Rhinitis, sore throat, cough	Clinical, nasal/throat swab if prolonged	Vitamin C daily decreases duration/severity in male swimmers	Supportive care, oral hydration, NSAIDs	Constantini	Spence	Dubnov [53]
		<30% cases from pathogen, i.e., EBV, adenovirus, rhinovirus	64% competitive swimmers aged 11–24							
	Ear infections (otitis externa)	Water with bacterial loads, especially high fecal indicator loads, hot weather	Swimmers, especially open water	Pruritis, ear canal edema, pain. Occlusion, lymphadenopathy, fever if severe	Clinical, otoscopic	Minimize manipulation or object insertion into ear	Antibiotic eardrops +/− hydrocortisone drops 3–4 times daily x 1 week, pain control with NSAIDs. Aural cleansing	Medina-Blasini		

	Rhinitis	Cold air, general allergen (dust, mites, etc.), emission pollutants	Athletes with atopy, asthma	Nasal obstruction, rhinorrhea, sneezing, nasal itching, eye discomfort, conjunctivitis	Clinical, patch testing if severe	Cromolyn several days +30 min prior to exposure, immunotherapy for 3 months prior if severe. Nasal strips for sleeping	2nd-gen antihistamines, Azelastine. Topical and oral decongestants[a], nasal sprays (saline, cromolyn, steroid), lubricating eye drops			
		NCl3	50% swimmers, up to 74% elite swimmers			Use of nose clip for 30 days during training		Harder [31]	Chamberlain [32]	Russo
			Intense swimming >10 h/week							
Gastrointestinal	Gastroenteritis	Water with high viral or bacterial loads, after heavy rainfall	Open water swimmers	Stomachache, fatigue, nausea, vomiting, diarrhea, dizziness, headache	Clinical based on timing of symptoms, monitor volume status	Exposure minimzation, training techniques to minimize ingestion, regular water indicator testing,	Supportive care, fluid and electrolyte replacement	Harder [31], Laurdison		

(continued)

Table 12.3 (continued)

System affected	Illnesses	Trigger	Athletes at risk	Details	Diagnosis	Prevention Strategy	Acute management	References		
		Foodborne pathogens	Athletes traveling, in shared living spaces, contact sports		Stool sample if refractory or outbreak. Laboratory studies if bloody stools, > 48 h of frequent loose stools, severe pain	Prophylactic bismuth QID	Pathogen-specific antimicrobial treatment if tested positive	Karageanes	Stellingwerff [52]	
						Travel with pre-packaged food, eat hot food only, avoid raw foods	Loperamide to decrease bowel movement frequency			
						Avoid tap water or ice in beverages, bring/ purchase bottled water or drink water boiled @ 100C for >5 min or halogenated				
Cardiovascular	Hypertension, increased heart rate variability	PM	All athletes	Asymptomatic or decreased performance, muscle cramping, chest tightness	Continuous heart rate, blood pressure monitoring	Exposure minimization, athletic filtration masks, antioxidant supplementation (ie fish oil, beetroot juice)	Early evidence suggests montelukast may reverse effects	Cutrufello [24]		Beijing 2012

	Angina, myocardial ischemia		Prior or family history of coronary artery disease	Chest/jaw/shoulder pain, shortness of breath, diaphoresis, stomachache, nausea, dizziness, sudden collapse	EKG, serum studies	Thorough screening, annual physicals	Nitrates, oxygen, aspirin	Giles [29]		
Dermatologic	Dermatitis	Chemical pollutants, aquatic microbes, and species-specific toxins			Clinical, Punch biopsy can confirm but not necessary	Avoidance, Shower/bathe immediately after water exposure with thorough scrubbing unless other injuries or sting injuries present	TCS (mid-high potency), topical and oral antihistamines, Oral prednisone[a] if severe. Sunscreen, unscented lotions			
	"Seabather's Eruption"	Cnidarian larvae (saltwater jellyfish)	Swimmers in the Atlantic Coastal region	Pruritis, fatigue, fever, possibly chills, nausea, headache, abdominal pain, diarrhea	Affects covered skin.		Additionally, alcohol, vinegar	Burnett 94, WHO [9]		
	Swimmer's itch	Schistosomes (fresh or saltwater)	Swimmers worldwide, more prevalent in summer	Pruritis, tingling, with papules or blisters	Affects exposed skin		Additionally, epsom solts, baking soda paste			

ICS inhaled corticosteroid, *TCS* topical corticosteroid, *IBA* inhaled beta-2 agonist, *SABA* short-acting beta agonist, *AHR* airway hyperresponsivity, *EIB* exercise-induced bronchospasm, *WADA* World Anti-Doping Agency

[a]Indicated possibly banned or regulated substance, should contact regulatory body and obtain special permission prior to event

Table 12.4 Nonpharmacologic strategies to manage exercise-induced bronchoconstriction

Strategy	Intervention	Potential effect	Evidence level	Pitfalls	References
Pre-exercise warm-up	Repetitive 30-s bouts close to Vo_{2max}/HR_{max}	Reduces postexercise fall in FEV_1	Good	May accumulate peripheral fatigue prior to exertion	[3, 4]
Face masks	HME face masks	Reduces postexercise fall in FEV_1	Insufficient	May affect ventilation and be associated with discomfort	[5–8]
Omega-3 fatty acid supplementation	3 g/d EPA and 2 g/d DHA	Reduces systemic inflammation	Medium	Side effects: Acid reflux, bloating, diarrhea, and nausea	[9–13]
		Reduces airway inflammation	Insufficient		
		Reduces postexercise fall in FEV_1	Insufficient		
Caffeine	5–10 mg/kg bw	Induces bronchodilation	Good	Slow absorption	[14–21]
		Reduces postexercise fall in FEV_1	Medium	Side effects: Muscle tremors, tachycardia	
		Improves respiratory muscle fatigue resilience	Medium		
		Counteracts exercise-induced hypoxemia	Insufficient		
Vitamins and antioxidants	1500 mg/d vitamin C	Reduces systemic inflammation	Medium	Side effects: diarrhea, vomiting, headache, insomnia, nausea, kidney stones	[22–33]
	64 mg/d b-carotene	Scavenge ROS	Medium		
		Reduces postexercise fall in FEV_1	Insufficient		
Breathing control	See Table 12.2	Reduces perception of respiratory symptoms	Insufficient		[34–37]
Respiratory muscle training	30 breaths × 2/d at 50% of MIP	Improves respiratory muscle fatigue resilience	Good	Time-consuming	[27, 38–52]
		Reduces asthma severity	Insufficient		

Reproduced with permission from Dickinson et al. [2]
Abbreviations: HR_{max} maximum heart rate, *MIP* maximum inspiratory pressure, *ROS* reactive oxygen species; VO_{2max} maximum oxygen consumption

12.6 Environmental Pollution Avoidance and Mitigation at the Time of Training or Competition

Coaches and physicians should be aware of how pollutants may fluctuate based on time of day, time of year, and location when professional testing methods are not available. Studies have shown that CO, NO_2, and PM 2.5 are relatively higher during the winter months, and ozone predominates in the warmer spring and summer months [27, 48]. Selecting venues for competition near parks instead of major roadways can reduce exposure to PM1 [27] and to NO_2 by almost 50%, according to one study in Beijing by Tao et al.; however, concentrations and inhalation of ozone at roadside areas were 25%

lower than in parks [48]. When jogging or cycling near a major highway, as little as 50 m of added distance from the highway can lead to an approximate threefold decrease in PM exposure, with some states advocating for a 100 m minimum distance for development of new athletic facilities [1, 24, 27]. Pollutants such as CO, NO_2, and PM 2.5 have been shown to paradoxically decrease from the hours of 4:00–6:00 PM; ozone levels are usually lowest from 6:00–8:00 AM and 6:00–8:00 PM [48]. Thus, morning training sessions are typically best, especially in the summer months when ozone levels are higher, as there are generally low pollutant concentrations due to minimal traffic and low solar radiation [3, 27, 29].

If an outdoor practice or competition is being held during a high pollution time frame, it is recommended to arrive to the venue just before the event begins [49]. Encourage athletic training staff to perform team warm-ups prior to transport to the venue or locate an indoor training room within the stadium or nearby campus to warm up in. For indoor venues, encourage electric-powered alternatives to gas or propane-powered machines like Zambonis in ice arenas or space heaters, which are known to cause significant PM production [24, 50]. All indoor spaces should have high-efficiency particulate air (HEPA) filters installed [40].

Face masks should be available to all athletes to be worn as tolerated by the athlete. Respiratory face masks, such as N95s, may filter out harmful pollutants and reduce pulmonary complications [2, 29]. Face masks also have cardiovascular health benefits during exercise including decreased blood pressure and increased pulmonary function and heart rate variability but have also been shown to occasionally cause mild shortness of breath and increase CO_2 retention and rebreathing [40, 51]. Given recent public opinion shift from the COVID-19 pandemic, mask wearing may be more acceptable socially and better tolerated by athletes [8, 51]. It is the job of coaches and team physicians to make sure the athletes are up to date on mask fittings and are educated on proper mask wearing protocols. Figure 12.3 summarizes important measures to take to minimize air pollution and its impacts on athletes.

While in season, and especially while traveling, athletes encounter a multitude of physiologic stressors that are exacerbated by local pollution. A strong emphasis should be placed on nutrition and supplementation to optimize athlete health and minimize effects of inflammation and ROS [52]. Athletes and coaches and trainers should collaborate with nutritionists to ensure adequate carbohydrate and antioxidant intake and may do so by bringing additional nonperishable foods and high-protein snacks like canned meals, protein bars, or nuts while on the road [52]. A Mediterranean diet consisting of foods high in antioxidants, omega-3 fatty acids, fiber, and protein and low in added sodium and sugar may help athletes mitigate airway inflammation and protect against tobacco smoke exposure, although this has not been proven to protect against other types of air pollution [12]. EPA and DHA supplementation (omega-3 fatty acids) has been found to reduce airway inflammation and bronchoconstriction in asthmatic patients to a similar extent as pharmacologic treatment with montelukast [2]. Supplementation of carotenoids in conjunction with fruit and vegetable intake has shown improved clinical asthma outcomes, although this has not been significantly proven in athletic populations [12, 52]. In swimmers, vitamin C supplementation 1 g/day has shown to improve upper respiratory tract infection (URI) duration and severity in males [53], and vitamin D 2000 IU/day improved duration of URI compared to placebo [53]. Prior to competing in high O_3 environments, 1 week of supplementation with 250–650 mg vitamin C, 75–100 mg vitamin E, and 25 mg of beta carotene may offer protective effects with minimal side effects [40]. Caffeine doses as low as 3 mg/kg of bodyweight have been shown to improve respiratory dynamics and endurance during submaximal exercise, protecting even elite-level athletes from desaturation [2]. Beta-alanine supplementation has been shown to increase carnosine concentration in fast and slow twitch muscle fibers, reducing reactive oxygen species and inflammation [54]. Creatine, frequently used by strength athletes, exacerbates inflammation and severity of bronchoconstriction in mice models, although this has not been proven to impact athletes [2].

Fig. 12.3 Infographic summary of strategies that may be taken to mitigate the effects of air pollution exposure on athletes

In preparation for aquatic competitions, incident response plans should be designed to prevent mass exposure to aquatic toxins. Rapid testing, including QPCR for certain bacterial species such as *Enterococcus*, has been shown to be a reliable predictor of pollution-related illness and should be conducted prior to, and on the day of, large-scale events [10]. Response plans should include additional sampling of waters after heavy rainfall, reported municipal waste treatment incidents, and in times of visible algae blooms or recent tide changes [9].

12.7 Strategies for Reducing Pollution Effects on Athletes and Spectators

The sports medicine physician's role is twofold: first, preparing to diagnose, treat, and monitor the pollution-related medical conditions as discussed above and, secondly, to participate in event planning and contingency planning. Most sources advocate for a clear plan made in advance, as well as a chain of command for decision-making regarding the safety of the athletes. Individuals should be assigned to monitoring weather conditions, air and water quality, and traffic conditions for large-scale events. Viana et al. demonstrated that sensing nodes deployed in major city stadiums could accurately track ambient air quality and measure ratios of different pollutants to provide insights into potential emission sources [55]. As pollution monitoring becomes more readily available, this data should be integrated into technological applications that can more accurately guide athletes and coaches when determining ideal locations and timing of training or competitions [8]. Certain athletic organizations are faced with greater logistical challenges in responding to pollution patterns. These include teams or athletes based in urban or rural areas, those with limited resources for travel or facilities upgrades, or sports that are not effectively trained in an indoor, pollution-controlled or in a controlled aquatic environment. It is the role of the public health officials, athletic directors, and coaches to collaborate with physicians in advance to anticipate and navigate scheduling conflicts that may arise when attempting to reschedule events, particularly in the case of large regional, national, or international sporting events.

12.8 Take-Home Pearls Regarding Environmental Pollutants

- Be aware of athletes' comorbidities, and consult a primary care provider or specialist before athletic participation if there are any concerns.
- Sideline kits and medical tents should have portable oxygen, IV electrolyte solutions, second-generation antihistamines, respirator masks, filtered water, topical corticosteroid creams, and athlete-specific medications, which should be cross-referenced against anti-doping agency banned or restricted medication lists.
- Ensure proper athlete hydration and nutrition, including antioxidant supplementation, to optimize endurance and metabolic response to pollutants.
- Limit outdoor training during times of high pollution and avoid proximity to major roadways, at least 100 m away if possible. Indoor environments should have HEPA-filtered air.
- Limit pre-competition warm-ups >15 min in high-pollution environments, as this may impair subsequent performance and worsen symptom severity.
- Test swimming pools and open waters for chemical and bacterial concentrations prior to and day of any competitions, and minimize excessive chlorination of pools.
- Monitor air pollution indexes and be familiar with common pollution trends in the area.
- Collaborate with local governments on policies reducing traffic and industrial emissions within the community, especially before and during large-scale events.

References

1. Rundell KW, Smoliga JM, Bougault V. Exercise-induced bronchoconstriction and the air we breathe. Immunol Allergy Clin N Am. 2018;38(2):183–204. https://doi.org/10.1016/j.iac.2018.01.009.
2. Dickinson J, Amirav I, Hostrup M. Nonpharmacologic strategies to manage exercise-induced bronchoconstriction. Immunol Allergy Clin N Am. 2018;38(2):245–58. https://doi.org/10.1016/j.iac.2018.01.012.
3. Marr LC, Ely MR. Effect of air pollution on marathon running performance. Med Sci Sports Exerc. 2010;42(3):585–91. https://doi.org/10.1249/MSS.0b013e3181b84a85.
4. Tolbert PE, Mulholland JA, MacIntosh DL, et al. Air quality and pediatric emergency room visits for asthma in Atlanta, Georgia, USA. Am J Epidemiol. 2000;151(8):798–810. https://doi.org/10.1093/OXFORDJOURNALS.AJE.A010280.
5. Madaniyazi L, Xerxes S. Outdoor air pollution and the onset and exacerbation of asthma. Chronic Dis Transl Med. 2021;7(2):100–6. https://doi.org/10.1016/J.CDTM.2021.04.003.
6. Li Y, Wang W, Kan H, Xu X, Chen B. Air quality and outpatient visits for asthma in adults during the 2008 Summer Olympic Games in Beijing. Sci Total Environ. 2010;408(5):1226–7. https://doi.org/10.1016/j.scitotenv.2009.11.035.
7. Silveira AC, Hasegawa JS, Cruz R, et al. Effects of air pollution exposure on inflammatory and endurance performance in recreationally trained cyclists adapted to traffic-related air pollution. Am J Physiol—Regul

Integr Comp Physiol. 2022;322(6):R562–70. https://doi.org/10.1152/ajpregu.00305.2021.
8. Araneda OF, Kosche-Cárcamo F, Verdugo-Marchese H, Tuesta M. Pulmonary effects due to physical exercise in polluted air: evidence from studies conducted on healthy humans. Appl Sci. 2021;11(7) https://doi.org/10.3390/app11072890.
9. World Health Organization. Guidelines for safe recreational water. Volume 1. Coastal and fresh waters. Geneva: World Health Organization; 2003. p. 219. http://apps.who.int/bookorders. Accessed 21 Mar 2024
10. Wade TJ, Calderon RL, Sams E, et al. Rapidly measured indicators of recreational water quality are predictive of swimming-associated gastrointestinal illness. Environ Health Perspect. 2006;114(1):24–8. https://doi.org/10.1289/ehp.8273.
11. Shephard RJ. Athletic performance and urban air pollution. Can Med Assoc J. 1984;131(2):105–9.
12. Whyand T, Hurst JR, Beckles M, Caplin ME. Pollution and respiratory disease: can diet or supplements help? A review. Respir Res. 2018;19(1) https://doi.org/10.1186/s12931-018-0785-0.
13. World Health Organization. World health statistics 2023: monitoring health for the SDGs, sustainable development goals. Geneva: World Health Organization; 2023.
14. Streets DG, Fu JS, Jang CJ, et al. Air quality during the 2008 Beijing Olympic Games. Atmos Environ. 2007;41(3):480–92. https://doi.org/10.1016/J.ATMOSENV.2006.08.046.
15. Wang M, Zhu T, Zheng J, et al. Use of a mobile laboratory to evaluate changes in on-road air pollutants during the Beijing 2008 summer olympics. Atmos Chem Phys. 2009;9(21):8247–63. https://doi.org/10.5194/acp-9-8247-2009.
16. Mijling B, Van Der ARJ, Boersma KF, Van Roozendael M, De Smedt I, Kelder HM. Reductions of NO_2 detected from space during the 2008 Beijing Olympic Games. Geophys Res Lett. 2009;36(13):1–6. https://doi.org/10.1029/2009GL038943.
17. Rich DQ, Kipen HM, Huang W, et al. Association between changes in air pollution levels during the Beijing olympics and biomarkers of inflammation and thrombosis in healthy young adults. JAMA. 2012;307(19):2068–78. https://doi.org/10.1001/jama.2012.3488.
18. Su C, Breitner S, Schneider A, et al. Short-term effects of fine particulate air pollution on cardiovascular hospital emergency room visits: a time-series study in Beijing, China. Int Arch Occup Environ Health. 2016;89(4):641–57. https://doi.org/10.1007/S00420-015-1102-6.
19. World Health Organization. WHO global air quality guidelines: particulate matter (PM2.5 and PM10), ozone, nitrogen dioxide, sulfur dioxide and carbon monoxide. In: Vol License: C. Geneva: World Health Organization; 2021. p. 7823–30. https://iris.who.int/handle/10665/345329.
20. Gauderman WJ, Avol E, Gilliland F, et al. The effect of air pollution on lung development from 10 to 18 years of age. N Engl J Med. 2004;351(11):1057–67. https://doi.org/10.1056/NEJMOA040610/SUPPL_FILE/1057SA1.PDF.
21. Rundell KW. Effect of air pollution on athlete health and performance. Br J Sports Med. 2012;46(6):407–12. https://doi.org/10.1136/bjsports-2011-090823.
22. Ferin J, Oberdörster G, Penney DP. Pulmonary retention of ultrafine and fine particles in rats. Am J Respir Cell Mol Biol. 1992;6(5):535–42. https://doi.org/10.1165/AJRCMB/6.5.535.
23. Weisenberg H, Zhao T, Heinrich J. Combinations of epidemiological and experimental studies in air pollution research: a narrative review. Int J Environ Res Public Health. 2020;17(2) https://doi.org/10.3390/IJERPH17020385.
24. Cutrufello PT, Smoliga JM, Rundell KW. Small things make a big difference: particulate matter and exercise. Sport Med. 2012;42(12):1041–58. https://doi.org/10.2165/11635170-000000000-00000.
25. Cutrufello PT, Rundell KW, Smoliga JM, Stylianides GA. Inhaled whole exhaust and its effect on exercise performance and vascular function. Inhal Toxicol. 2011;23(11):658–67. https://doi.org/10.3109/08958378.2011.604106.
26. Rundell KW, Hoffman JR, Caviston R, Bulbulian R, Hollenbach AM. Inhalation of ultrafine and fine particulate matter disrupts systemic vascular function. Inhal Toxicol. 2007;19(2):133–40. https://doi.org/10.1080/08958370601051727.
27. Rundell K, Caviston R, Hollenbach A, Murphy K. Vehicular air pollution, playgrounds, and youth athletic fields. Inhal Toxicol. 2006;18(8):541–7. https://doi.org/10.1080/08958370600685640.
28. Rundell KW. Pulmonary function decay in women ice hockey players: is there a relationship to ice rink quality? Inhal Toxicol. 2004;16(3):117–23. https://doi.org/10.1080/08958370490270918.
29. Giles LV, Koehle MS. The health effects of exercising in air pollution. Sport Med. 2014;44(2):223–49. https://doi.org/10.1007/s40279-013-0108-z.
30. Koike A, Wasserman K, Armon Y, Weiler-Ravell D. The work-rate-dependent effect of carbon monoxide on ventilatory control during exercise. Respir Physiol. 1991;85(2):169–83. https://doi.org/10.1016/0034-5687(91)90059-R.
31. Harder-Lauridsen NM, Kuhn KG, Erichsen AC, Mølbak K, Ethelberg S. Gastrointestinal illness among triathletes swimming in non-polluted versus polluted seawater affected by heavy rainfall, Denmark, 2010–2011. PLoS One. 2013;8(11):2010–1. https://doi.org/10.1371/journal.pone.0078371.
32. Chamberlain M, Marshall AN, Keeler S. Open water swimming: medical and water quality considerations. Curr Sports Med Rep. 2019;18(4):121–8. https://doi.org/10.1249/JSR.0000000000000582.

33. World Health Organization. Guidelines for safe recreational water environments: volume 2. Swimming pools and similar environments. World Health Organ. 2006;2:146.
34. Lichter A, Pestel N, Sommer E. Productivity effects of air pollution: evidence from professional soccer. Labour Econ. 2017;48:54–66. https://doi.org/10.1016/J.LABECO.2017.06.002.
35. Beavan A, Härtel S, Spielmann J, Koehle M. Air pollution and elite adolescent soccer players' performance and well-being; an observational study. Environ Int. 2023;175 https://doi.org/10.1016/j.envint.2023.107943.
36. Beavan A, Härtel S, Spielmann J, Koehle M. Air pollution, a worthy opponent? How pollution levels impair athlete performance across physical, technical, and cognitive domains. Sci Total Environ. 2023;900 https://doi.org/10.1016/j.scitotenv.2023.165707.
37. Heintz EC, Scott DP, Simms KR, Foreman JJ. Air quality is predictive of mistakes in professional baseball and American football. Int J Environ Res Public Health. 2023;20(1) https://doi.org/10.3390/ijerph20010542.
38. Cusick M, Rowland ST, DeFelice N. Impact of air pollution on running performance. Sci Rep. 2023;13(1):1–9. https://doi.org/10.1038/s41598-023-28802-x.
39. Bos I, De Boever P, Int Panis L, Meeusen R. Physical activity, air pollution and the brain. Sport Med. 2014;44(11):1505–18. https://doi.org/10.1007/s40279-014-0222-6.
40. Hung A, Koch S, Bougault V, et al. Personal strategies to mitigate the effects of air pollution exposure during sport and exercise: a narrative review and position statement by the Canadian Academy of Sport and Exercise Medicine and the Canadian Society for Exercise Physiology. Br J Sports Med. 2023;57(4):193–202. https://doi.org/10.1136/bjsports-2022-106161.
41. Wong CM, Ou CQ, Thach TQ, et al. Does regular exercise protect against air pollution-associated mortality? Prev Med (Baltim). 2007;44(5):386–92. https://doi.org/10.1016/J.YPMED.2006.12.012.
42. Ora J, De Marco P, Gabriele M, Cazzola M, Rogliani P. Exercise-induced asthma: managing respiratory issues in athletes. J Funct Morphol Kinesiol. 2024;9(1):15. https://doi.org/10.3390/jfmk9010015.
43. Allado E, Poussel M, Hily O, Chenuel B. The interest of rehabilitation of respiratory disorders in athletes: myth or reality? Ann Phys Rehabil Med. 2022;65(4) https://doi.org/10.1016/j.rehab.2020.101461.
44. Boskabady M, Esmaeilizadeh M, Boskabady M. The effect of exposure to chlorine on pulmonary function tests and respiratory and allergic symptoms in Iranian lifeguards. Toxicol Ind Health. 2012;30(3):218–24. https://doi.org/10.1177/0748233712454465.
45. Päivinen M, Keskinen K, Tikkanen H. Swimming-induced changes in pulmonary function:special observations for clinical testing. BMC Sports Sci Med Rehabil. 2021;13(1) https://doi.org/10.1186/S13102-021-00277-1.
46. Price OJ, Hull JH, Backer V, Hostrup M, Ansley L. The impact of exercise-induced bronchoconstriction on athletic performance: a systematic review. Sport Med. 2014;44(12):1749–61. https://doi.org/10.1007/s40279-014-0238-y.
47. McKenzie DC, Fitch KD. The asthmatic athlete: inhaled beta-2 agonists, sport performance, and doping. Clin J Sport Med. 2011;21(1):46–50. https://doi.org/10.1097/JAE.0b013e318203c0ef.
48. Tao SC, Gao SH, Xiong XZ, Yao JL, Deng SX. Air pollution characteristics and jogger inhalation exposure in typical running area of Beijing. Huan jing ke xue= Huanjing kexue. 2018;39(8):3580–90. https://doi.org/10.13227/J.HJKX.201711178.
49. Cressey D, Callaway E. Science at the Olympics: team science. Nature. 2012;487(7407):290–2. https://doi.org/10.1038/487290a.
50. Rundell KW, Spiering BA, Baumann JM, Evans TM. Bronchoconstriction provoked by exercise in a high-particulate-matter environment is attenuated by montelukast. Inhal Toxicol. 2005;17(2):99–105. https://doi.org/10.1080/08958370590899479.
51. Epstein D, Korytny A, Isenberg Y, et al. Return to training in the COVID-19 era: the physiological effects of face masks during exercise. Scand J Med Sci Sport. 2021;31(1):70–5. https://doi.org/10.1111/sms.13832.
52. Stellingwerff T, Pyne DB, Burke LM. Nutrition considerations in special environments for aquatic sports. Int J Sport Nutr Exerc Metab. 2014;24(4):470–9. https://doi.org/10.1123/ijsnem.2014-0014.
53. Dubnov-Raz G, Rinat B, Hemilä H, Choleva L, Cohen AH, Constantini NW. Vitamin D supplementation and upper respiratory tract infections in adolescent swimmers: a randomized controlled trial. Pediatr Exerc Sci. 2015;27(1):113–9. https://doi.org/10.1123/pes.2014-0030.
54. Hostrup M, Bangsbo J. Improving beta-alanine supplementation strategy to enhance exercise performance in athletes. J Physiol. 2016;594(17):4701–2. https://doi.org/10.1113/JP272530.
55. Viana M, Karatzas K, Arvanitis A, et al. Air quality sensors systems as tools to support guidance in athletics stadia for elite and recreational athletes. Int J Environ Res Public Health. 2022;19(6) https://doi.org/10.3390/ijerph19063561.